VAUXHALL
ASTRA & BELMONT
1980 to October 1991

All Vauxhall Astra, Belmont and Astra Belmont models except Astra/Astramax Van and Diesel engine models. Does not cover revised Astra range introduced in October 1991.

(1818 – IWI)

A Haynes Handbook and Drivers Guide
© Haynes Publishing 1994

Printed and published by
J H Haynes & Co Ltd Sparkford
Nr Yeovil Somerset BA22 7JJ
England

ISBN 1 85010 818 8

British Library Cataloguing in Publication Data. A catalogue record for this book is available from the British Library

All rights reserved. No part of this book may be reproduced or transmitted in any form or by any means, electronic or mechanical, including photocopying, recording or by any information storage or retrieval system, without permission in writing from the copyright holder

HANDBOOK & DRIVERS GUIDE
by A K Legg LAE MIMI

Full details of all servicing and repair tasks for the models covered by this Handbook can be found in the relevant Owners Workshop Manual – OWM **635** Astra 1980 to October 1984, or OWM **1136** Astra & Belmont October 1984 to October 1991.

ABCDEFGHJKLMNOPQRST

ACKNOWLEDGEMENTS

We gratefully acknowledge the assistance of RoSPA in compiling the information used in this Handbook. Thanks are due to Duckhams Oils who provided lubrication data. Certain other illustrations are the copyright of Vauxhall Motors Ltd., and are used with their permission. Additional photographs were supplied by Quadrant Picture Library/Auto Express. Thanks are also due to all those people at Sparkford who helped in the production of this Handbook.

We take great pride in the accuracy of information given in this Handbook, but vehicle manufacturers make alterations and design changes during the production run of a particular vehicle of which they do not inform us. No liability can be accepted by the authors or publishers for loss, damage or injury caused by any errors in, or omissions from, the information given.

CONTENTS

About this Handbook	5
The Astra and Belmont family	7
Buying and selling	9
General tips on buying	9
Points to look for when buying	10
General tips on selling	11
Dimensions and weights	12
Controls and equipment	15
Driver's instruments and controls	20
Interior equipment	30
Exterior equipment	36
Accidents and emergencies	39
How to cope with an accident	39
First aid	40
Requirements of the law	42
Essential details to record	42
Accident report form	43
How to cope with a fire	46
How to cope with a broken windscreen	46
What to do if your car is broken into	46
Breakdowns	47
Breakdowns on an ordinary road	47
Motorway breakdowns	48
Changing a wheel	48
Towing	51
Starting a car with a flat battery	52
What to carry in case of a breakdown	53
Driving safety	55
Before starting a journey	55
Driving in bad weather	56
Motorway driving	57
Towing a trailer or caravan	58
Alcohol and driving	60
Skid control	60
Advice to women drivers	61
Child safety	63
Driving abroad	65
Reducing the cost of motoring	69
Car crime prevention	71
Service specifications	73
Regular checks	77
Checking oil level	80
Checking coolant level	80
Checking brake fluid level	81
Checking tyres	82
Checking washer fluid level	83
Checking battery electrolyte level	84
Checking wipers and washers	85
Checking lights and horn	86
Checking for fluid leaks	87
Servicing	89
Service schedule	90
Safety first!	98
Buying spare parts	101
Service tasks	106
Seasonal servicing	125
Tools	128
What to buy	128
Care of your tools	128
Bodywork and interior care	129
Cleaning the interior	129
Cleaning the bodywork	129
Dealing with scratches	130
Bulb, fuse and relay renewal	131
Bulbs – exterior lights	132
Bulbs – interior lights	138
Fuses	139
Relays	140
Preparing for the MOT test	141
Fault finding	143
Car jargon	149
Local radio frequencies	159
Conversion factors	164
Distance tables	166
Index	168
Servicing notes	172

VAUXHALL ASTRA & BELMONT

4 ABOUT THIS HANDBOOK

VAUXHALL ASTRA & BELMONT

ABOUT THIS HANDBOOK

The idea behind this Handbook is to help you to get the most out of your motoring.

Apart from the things that every owner needs to know, to deal with unexpected mishaps like a puncture or a blown light bulb, you'll find clearly-presented information on road safety, hints on driving abroad, and tips on how to prepare your car for the MOT test. We've also included details of local radio frequencies to help you avoid those inevitable 'jams', and there's a Section on what to do in the unfortunate event of an accident.

For those not familiar with their car, there's a Section to explain the location and the operation of the various controls and instruments. Additionally, there's a Section on fault finding, and a useful glossary of car jargon.

Garage labour charges usually form the major part of any car servicing bill, and we hope to help you to reduce those bills by carrying out the more straightforward routine servicing jobs yourself. If you're about to start carrying out your own servicing for the first time, we aim to provide you with easy-to-follow instructions, enabling you to carry out the simpler tasks which perhaps you've left to a garage or a 'car-minded' friend in the past. Even if you prefer to leave regular servicing to a suitably-qualified expert, by using this book you'll be able to carry out regular checks on your car, to make sure that your motoring is safe and hopefully trouble-free. You'll also find advice on buying suitable tools, and safety in the home workshop.

Some readers of this Handbook may not yet have bought a Vauxhall Astra or Belmont, so we've included a brief history of the range, and some useful tips on buying and selling.

All in all, we hope that this book will prove a handy companion for your motoring adventures, and hopefully we'll help to reduce the problems which inevitably crop up in everyday driving. If you're bitten by the DIY bug, and you're keen to tackle some of the more advanced repair jobs on your car, then you'll need our **Owners Workshop Manual** for your particular model (OWM 635 for 'Mark 1' Astra models manufactured up to September 1984, or OWM 1136 for 'Mark 2' Astra and Belmont models manufactured from October 1984 onwards). These manuals give a step-by-step guide to all the repair and overhaul tasks, with plenty of illustrations to make things even clearer.

Happy motoring!

VAUXHALL ASTRA & BELMONT

6 THE ASTRA & BELMONT FAMILY

▲ *Astra 1.3 GL Hatchback (1981)*

▲ *Astra 1.6 L Hatchback (1986)*

THE ASTRA & BELMONT FAMILY

The Vauxhall Astra was introduced to the UK market early in 1980, and in the field of middle-sized front-wheel-drive cars, it has gained a good reputation for having a reliable engine and rust-free bodywork.

Initially, the Astra was available in 5-door Hatchback or Estate form, having a 1297 cc engine and 4-speed gearbox, but it was not long before additional engines were introduced together with a 5-speed gearbox.

The most significant change to the body style occurred in October 1984 when the 'Mark 2' Astra was introduced. Along with other manufacturer's models of this period, the 'Mark 2' Astra has a more aerodynamic look about it, with 'rounded' body panels and glass. Following this, the Belmont Saloon version came on the scene early in 1986, being virtually the same as the Astra except for the rear end styling which incorporates a 'proper' boot. Mechanically, the 'Mark 2' Astra is very similar to its predecessor.

Engines now range from 1.2 litres to 2.0 litres in size, and include a twin-camshaft 16-valve version fitted to the Astra GTE. A large number of different trim levels have been available over the years, from the Base Hatchback, Saloon and Estate, to the luxury CD and performance-orientated SRi and GTE models. A wide range of standard and optional equipment is available across the model ranges, including items such as sunroofs, electric windows, central door locking and, on certain later models, anti-lock brakes.

All members of the Astra/Belmont family share a similar mechanical layout, and they all use the same large family of engines and gearboxes, although many detailed modifications and improvements have been made since the original design was introduced. On all models, the 4-cylinder in-line engine is mounted transversely at the front of the car, with the gearbox positioned on the left-hand end of the engine (when viewed from the driver's seat).

Most of the mechanical components of the cars are fairly conventional, and the main systems are designed to keep servicing time and costs as low as possible. The electronic ignition system (fitted to later models), clutch, brakes, wheel bearings, etc are all routine-maintenance-free, and should only require attention when certain items need to be renewed.

There should be no problem obtaining spare parts for these cars, so they are an ideal proposition for the enthusiastic amateur mechanic who wants to keep running costs to a minimum.

▲ *Astra 1.3 L Estate (1980)*

VAUXHALL ASTRA & BELMONT

8 THE ASTRA & BELMONT FAMILY

▲ *Astra Belmont 1.4 LX Saloon (1989)*

▲ *Astra GTE 2.0 (1987)*

BUYING AND SELLING

GENERAL TIPS ON BUYING
- Don't rush out and buy the first car to catch your attention
- Always buy from a recognised dealer in preference to buying privately
- Have the car checked over by someone knowledgeable before you buy it
- In general, you get what you pay for!

Before buying a second-hand car, it's worthwhile doing some homework to try and avoid some of the pitfalls waiting for the unwary. First of all, don't rush out and buy the first car to catch your attention (all that glitters is not gold!), and remember that much of the responsibility is yours when it comes to the soundness of the deal, especially when buying a car privately.

Wherever possible, buy from a recognised dealer, and check that the dealer is a member of the Retail Motor Industry Federation, as this will provide you with certain legal safeguards if you have any problems. If you buy from a dealer, you are covered by the Sale Of Goods Act, which in summary states that the goods must be fit for their intended purpose, the goods must be of proper quality, and the goods must be as described by the seller. If you're buying a car privately, ask to see the service receipts and the MOT certificates going back as long as the car has been in the possession of the current owner (this will help to establish that the car hasn't been stolen, and that the recorded mileage is genuine), and always ask to view the car at the seller's private address (to make sure that the car isn't being sold by an unscrupulous dealer posing as a private seller). Check the vehicle documents for obvious signs of forgery and, if in doubt, contact the DVLA and give them details of the Registration Document, as they will be able to run a check on its authenticity.

As far as the soundness of the car itself is concerned, a genuine service history is helpful. This is provided by the service book supplied with the car when new, which should be completed and officially stamped by an authorised garage after each service. Cars with a full service history (fsh) usually command a higher price than those without.

To check the condition of a car, a professional examination is well worthwhile, if you can afford it, and organisations such as the AA and RAC will be able to provide such a

BUYING AND SELLING

service. Otherwise, you must trust your own judgement, and/or that of a knowledgeable friend. Although it's tempting, try not to overlook the mechanical soundness of the car in favour of the overall appearance. It's relatively easy to clean and polish a car every week, but when were the brakes and tyres last checked? Above all, safety must always take priority. Don't view a car in the wet, as water on the bodywork can give a misleading impression of the condition of the paintwork. First of all, check around the outside of the car for rust, and for obvious signs of new or mismatched paintwork which might show that the car has been involved in an accident. Check the tyres for signs of unusual wear or damage, and check that the car 'sits' evenly on its suspension, with all four corners at a similar height. Open the bonnet and check for any obvious signs of fluid leakage (oil, water, brake fluid), then start the engine and listen for any unusual noises – some background noise is to be expected on older cars, but there should be no sinister rattles or bangs! Also listen to the exhaust to make sure that it isn't 'blowing' indicating the need for renewal, and check for signs of excessive exhaust smoke. Black smoke may be caused by poorly-adjusted fuel mixture, which can usually be rectified fairly easily, but blue smoke usually indicates worn engine components, which may prove expensive to repair. Finally, drive the car, and test the brakes, steering and gearbox. Make sure that the car doesn't pull to one side, and check that the steering feels positive and that the gears can be selected satisfactorily, without undue harshness or noise. Listen for any unusual noises or vibration, and keep an eye on the instruments and warning lights, to make sure that they are working and indicating correctly.

If all proves satisfactory, try to negotiate a suitable deal, but remember that a dealer has to work to a profit margin, and it's unlikely that you'll find a good car for a silly price. Always obtain a receipt for your money.

On the whole, it's true to say that you get what you pay for. Above all, don't be rushed into making a hasty decision.

POINTS TO LOOK FOR WHEN BUYING A VAUXHALL ASTRA OR BELMONT

In addition to the points outlined previously, there a few special points to look out for when buying an Astra or Belmont.

In general, the Astra/Belmont is a sound car with above-average reliability, and it has an excellent engine and gearbox. The 'Mark 2' introduced in October 1984 is generally a better buy than the 'Mark 1,' mainly because of its improved looks – in terms of reliability and mechanical design it is very similar to the previous version.

The bodywork is not prone to rust or paintwork problems; the upholstery is hard-wearing and should not show any serious signs of deterioration, unless the car has covered a very high mileage or has been mistreated.

Where the engine has covered in excess of 50 000 miles (80 000 km) or so, there are a few things worth checking which could avoid unforeseen expense after having acquired the car. The main problem is camshaft wear on the overhead camshaft engine (there is no similar problem with the overhead valve 1.2 litre engine). To check for this, remove the engine oil filler cap (with the engine stopped) and look inside – you will see a shiny part of the camshaft called a lobe. There are eight of these lobes on the camshaft, and their purpose is to open and close the valves one after the other. The camshaft lobes press down onto followers which in turn open the valves, and these followers are also prone to wear. Run a fingernail across the lobe which is visible through the oil filler cap hole, and check for ridges or scoring which would indicate excessive wear. Don't assume that the rest of the lobes are necessarily in good condition if the checked one is, as it is quite feasible for one of the other ones to be worn, but this check will give you an idea of the general wear present. If the lobes are worn, this reveals itself as a clattering noise from the top of the engine. To check for this, run the engine over various speeds while listening with the bonnet open; if the camshaft is worn, the clattering noise will increase and decrease according to the engine speed, and will be particularly loud at high engine speeds. Fitting a new camshaft is quite an expensive job, and allowance for this should be made when negotiating the price. However, once the job has been done, the engine should be good for another 50 000 miles, provided that the rest of the engine is in good condition.

On all carburettor engines, check for oil leaking from the bottom of the fuel pump, which would indicate a faulty pump internal seal. If this is evident, all that is required is to fit a new pump – this should not be expensive.

If the engine has covered a high mileage, check for possible high oil consumption which may point towards an engine overhaul. With the engine warmed up, rev the engine several times and check the colour of the exhaust – if the engine is using oil, the exhaust will be blue in colour. This does not necessarily mean that the engine must be completely overhauled, since the valve stem rubber oil seals may be causing the trouble, and this would only call for the cylinder head to be removed. The rubber oil seals become hardened with age and do not seal properly, and as a result oil is drawn into the engine and burnt.

An unusual feature of the Astra/Belmont is the design of the clutch, which allows the friction plate to be renewed without having to remove either the engine or the transmission. This makes the renewal of the clutch a quick and comparatively inexpensive repair.

When trying out the car, a check should be made for worn brake discs and shock absorbers. To do this, make an emergency stop by braking hard, and check if any vibration is felt through the steering wheel, which would indicate worn brake pads and/or discs. Worn brake pads would be confirmed if there is a grinding noise at the same time. While making the emergency stop, check also if the front of the car dips excessively, and whether it bounces up and down when actually stopped – this would indicate that the front shock absorbers are worn.

A good secondhand Astra should not be difficult to find, although the sports and Belmont versions may require a little more hunting. If you are looking for one of the larger-engined models, 2.0 litre models generally make better buys than 1.8 litre models.

GENERAL TIPS ON SELLING

- Make sure that the car is clean and tidy
- Make sure that all of the service documents, registration document etc, are available for inspection
- Ask yourself ... 'Would I buy this car?'

Obviously when selling a car, bear in mind the points which the prospective buyer should be looking for, as described in the above sections. It goes without saying that the car should be clean and tidy, as first impressions are important. Any fluid leaks should be cured, and there's no point in trying to disguise any major bodywork or mechanical problems.

If you're trading the car in with a dealer, you will always get a lower price than if you sell privately, but you can be fairly sure that there will be less comeback to you should any unexpected problems develop. If selling privately, don't allow the buyer to take the car away until you have his/her money, and it's a good idea to ask him/her to sign a piece of paper to say that he/she is happy to buy the car as viewed, just in case any problems develop later on. Give a receipt for the money paid.

DIMENSIONS AND WEIGHTS

Note: All figures are approximate, and will vary depending on model

DIMENSIONS [mm (in)]
Overall length

Hatchback	**3998** (157.5)
Belmont	**4218** (166.2)
'Mark 1' Estate	**4206** (165.7)
'Mark 2' Estate	**4228** (166.6)

Overall width

All 'Mark 1' models except SR	**1636** (64.5)
'Mark 1' SR models	**1656** (65.2)
'Mark 2' Hatchback	**1663** (65.5)
'Mark 2' Estate	**1666** (65.6)
Belmont	**1658** (65.3)

VAUXHALL ASTRA & BELMONT

DIMENSIONS AND WEIGHTS

Note: All figures are approximate, and will vary depending on model

Dimensions continued [mm (in)]

Overall height

'Mark 1' Hatchback	**1380** (54.4)
'Mark 2' Hatchback (except GTE)	**1400** (55.2)
'Mark 2' Astra GTE	**1395** (55.0)
'Mark 1' Estate	**1400** (55.2)
'Mark 2' Estate	**1430** (56.3)
Belmont	**1400** (55.2)

WEIGHTS [kg (lb)]

Nominal kerb weight

Hatchback	**830** to **975** (1830 to 2150)
Belmont	**890** to **1002** (1962 to 2209)
Estate	**895** to **1045** (1973 to 2304)
Convertible	**970** to **1045** (2139 to 2304)

Maximum roof rack load

All models	**80** (176)

Maximum towing weight (for trailer or caravan)

1.2 litre	**600** (1323)
1.3 litre manual	**900** to **950** (1985 to 2095)
1.3 litre automatic	**700** to **750** (1544 to 1654)
1.6 litre	**1000** (2205)
1.8 litre	**1000** (2205)
2.0 litre	**1200** (2646)

Maximum trailer/caravan noseweight

All models	**50** (110)

VAUXHALL ASTRA & BELMONT

14 CONTROLS AND EQUIPMENT

VAUXHALL ASTRA & BELMONT

CONTROLS AND EQUIPMENT

For those not familiar with the Vauxhall Astra and Belmont models, this Section will help to identify the instruments and controls. Typical instrument panel layouts are shown in the accompanying illustrations. The operation of most equipment is self-explanatory, but some items require further explanation to ensure that their use is fully understood.

Note that not all items are fitted to all models.

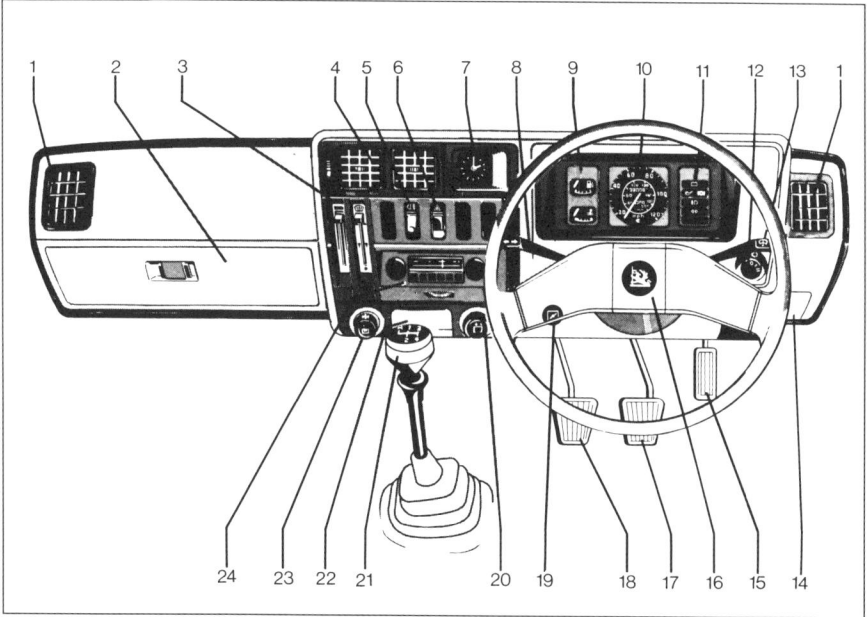

▲ Typical instrument panel layout for 'Mark 1' Astra models (up to and including 1984)

1 Demister vents for front door windows
2 Glovebox
3 Heater controls
4 Central air vents
5 Rear foglight switch
6 Hazard warning flasher switch
7 Clock
8 Direction indicator/headlamp dip/headlamp flasher stalk
9 Fuel gauge and engine coolant temperature gauge
10 Speedometer/mileometer/trip recorder
11 Warning light cluster
12 Lighting switch
13 Windscreen/tailgate/headlamp wash/wipe stalk
14 Fusebox
15 Accelerator pedal
16 Horn push
17 Brake pedal
18 Clutch pedal
19 Manual choke control (when fitted)
20 Cigarette lighter
21 Gearshift lever
22 Ashtray
23 Heater fan/heated rear window switch
24 Radio

VAUXHALL ASTRA & BELMONT

CONTROLS AND EQUIPMENT

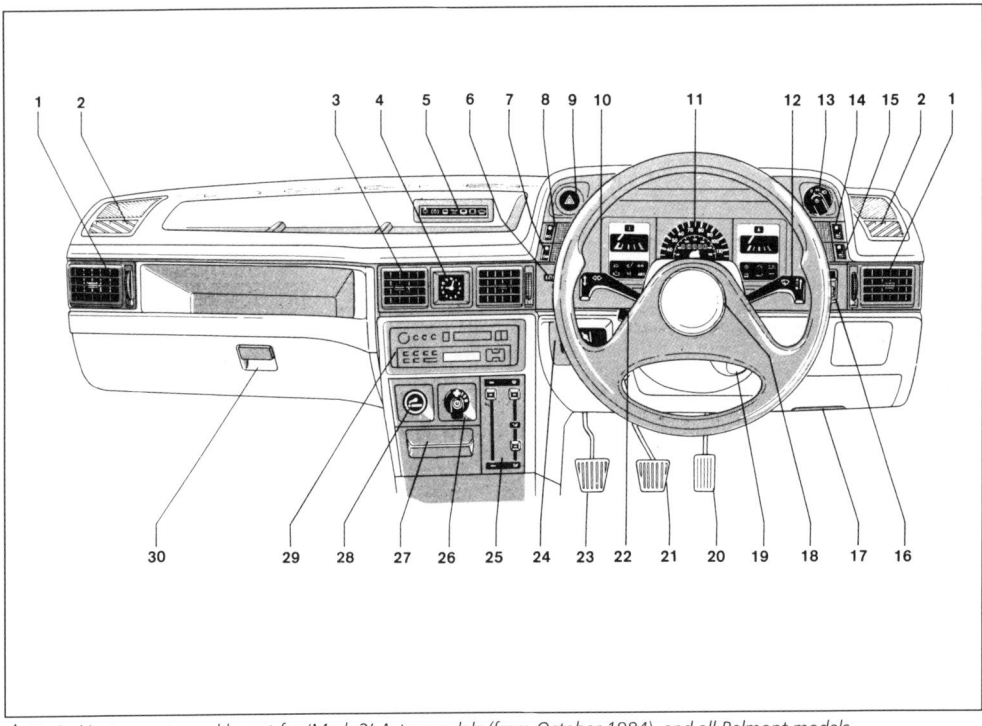

▲ *Typical instrument panel layout for 'Mark 2' Astra models (from October 1984), and all Belmont models*

1. Side air vents
2. Front door window demister vents
3. Central air vents
4. Clock
5. Check control display
6. Instrument panel illumination brightness control
7. Spare switch position
8. Spare switch position
9. Hazard warning flasher switch
10. Direction indicator/headlamp dip/headlamp flasher stalk
11. Instrument panel
12. Windscreen/tailgate/headlamp wash/wipe stalk
13. Lighting switch
14. Front foglight switch
15. Rear foglight switch
16. Headlamp beam adjustment control
17. Fusebox
18. Horn push
19. Ignition switch (hidden behind steering wheel)
20. Accelerator pedal
21. Brake pedal
22. Tiltable steering column adjustment
23. Clutch pedal
24. Manual choke control (when fitted)
25. Heater controls
26. Heater fan/heated rear window switch
27. Ashtray
28. Cigarette lighter
29. Radio
30. Glovebox

CONTROLS AND EQUIPMENT   17

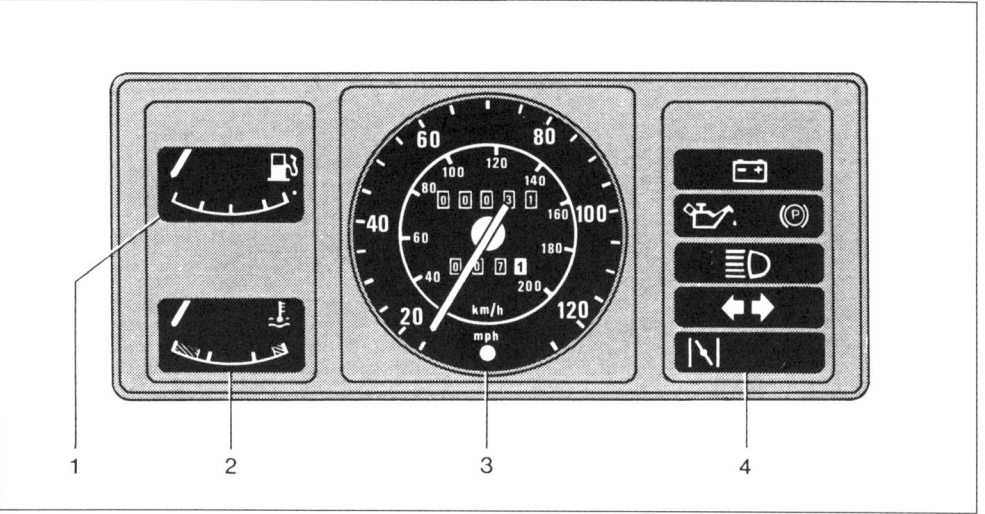

▲ Typical instrument panel on 'Mark 1' low series models

1 Fuel gauge
2 Engine coolant temperature gauge
3 Speedometer with mileometer and trip recorder
4 Warning lights for alternator/oil pressure/brake fluid level/headlight main beam/direction indicators/choke/trailer direction indicators

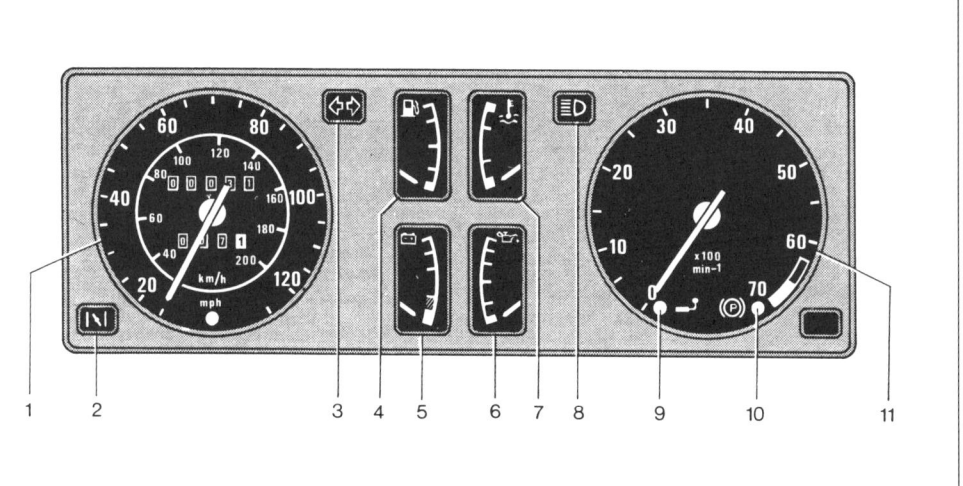

▲ Typical instrument panel on 'Mark 1' high series models

1 Speedometer
2 Manual choke warning light
3 Direction indicators warning light
4 Fuel gauge
5 Voltmeter with warning light for alternator
6 Oil pressure gauge with warning light for oil pressure
7 Engine coolant temperature gauge
8 Headlamp main beam warning light
9 Trailer direction indicator warning light
10 Handbrake warning light
11 Tachometer ('rev. counter')

VAUXHALL ASTRA & BELMONT

CONTROLS AND EQUIPMENT

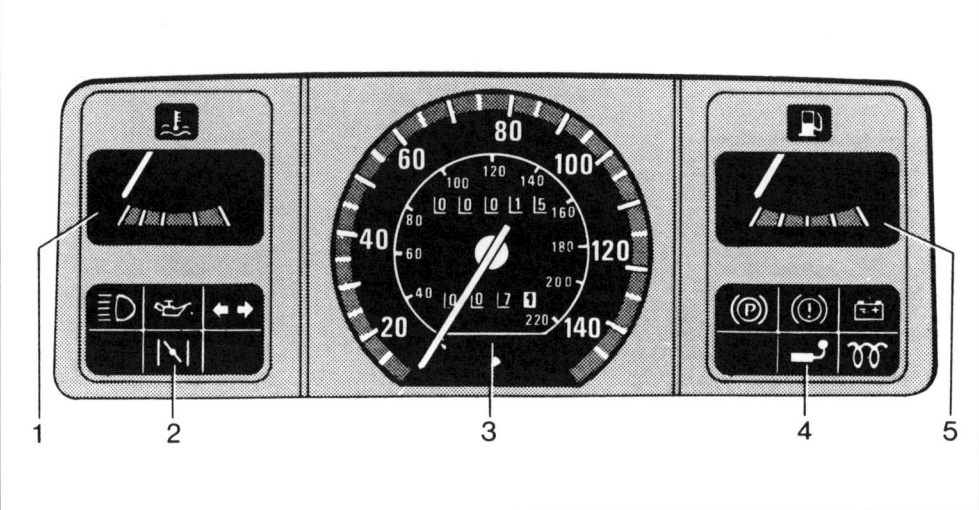

▲ Typical instrument panel on 'Mark 2' and Belmont low series models

1. Engine coolant temperature gauge
2. Warning lights for headlamp main beam/oil pressure/direction indicators/choke
3. Speedometer with mileometer and trip recorder
4. Warning lights for brake fluid level/handbrake/alternator/trailer direction indicators
5. Fuel gauge

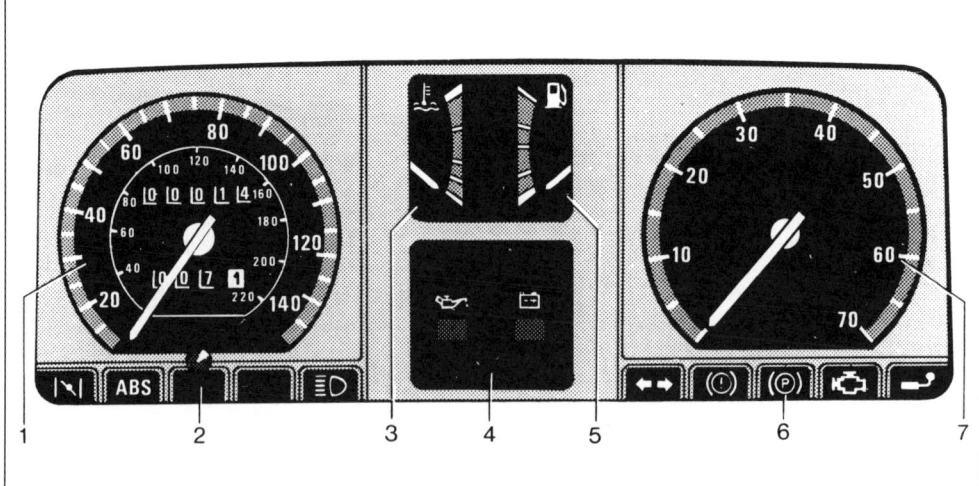

▲ Typical instrument panel on 'Mark 2' and Belmont high series models

1. Speedometer with mileometer and trip recorder
2. Warning lights for choke/ABS/headlamp main beam
3. Engine coolant temperature gauge
4. Warning lights for oil pressure/alternator
5. Fuel gauge with warning light
6. Warning lights for direction indicators/brake fluid level/engine control/trailer direction indicators
7. Tachometer ('rev. counter')

CONTROLS AND EQUIPMENT 19

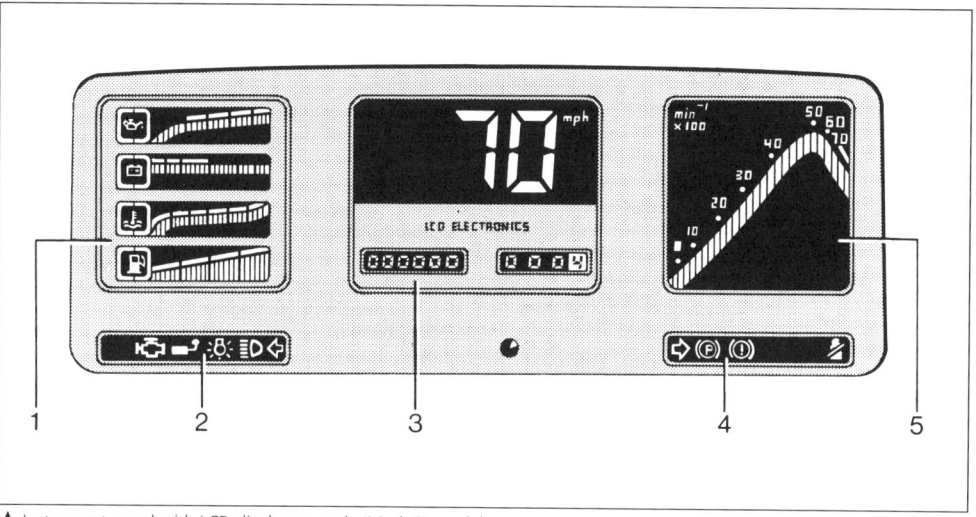

▲ Instrument panel with LCD display on early 'Mark 2' models

1. Oil pressure gauge/voltmeter/coolant temperature gauge/fuel gauge
2. Warning lights for engine control/trailer direction indicators/exterior lights/headlight main beam/left-hand direction indicator
3. Speedometer with mileometer, trip recorder and mph/kph selector
4. Warning lights for right-hand direction indicator/handbrake on/brake fluid level/seat belt latch
5. Tachometer ('rev. counter')

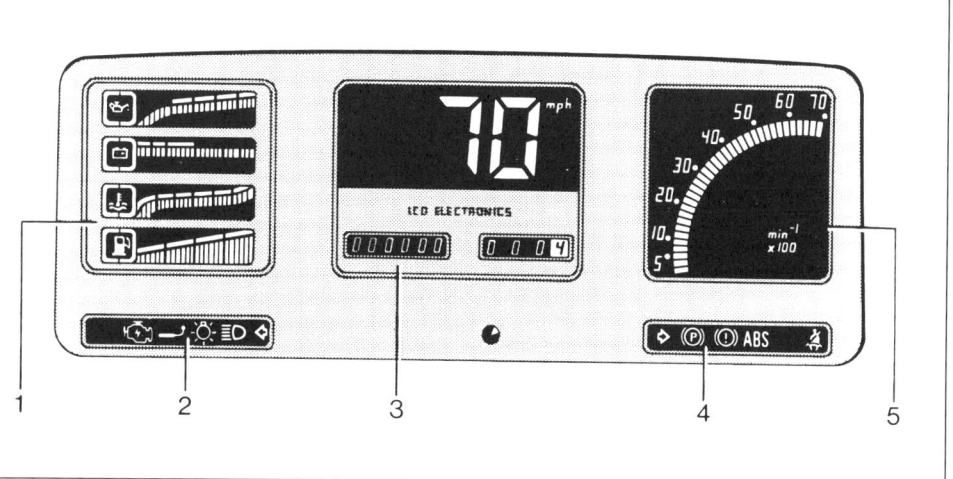

▲ Instrument panel with LCD display on later 'Mark 2' models

1. Oil pressure gauge/voltmeter/coolant temperature gauge/fuel gauge (all with warning zone)
2. Warning lights for engine control/trailer indicators/exterior lamps/headlamp main beam/left-hand direction indicator
3. Speedometer with mileometer, trip recorder and mph/kph selector
4. Warning lights for right-hand direction indicator/handbrake on/brake fluid level/ABS/seat belt
5. Tachometer ('rev. counter')

AUXHALL ASTRA & BELMONT

CONTROLS AND EQUIPMENT

DRIVER'S INSTRUMENTS AND CONTROLS

Speedometer

Indicates the car's road speed, and incorporates a mileage recorder. All models have a trip meter, which can be reset by pressing the button protruding from the instrument. On models with an LCD instrument panel display, the digital display can be changed from mph to kph by turning the trip meter reset knob.

Tachometer ('rev. counter')

Indicates engine speed in revolutions per minute (rpm or rev/min). For normal driving and the best fuel economy, the engine speed should be kept in the black zone – 3000 to 4500 rpm for 1.2 to 1.6 litre engines, and 3000 to 5000 rpm for 1.8 and 2.0 litre engines. The red outlined zone indicates engine speeds which should only be used for brief periods. The red zone indicates the danger zone, and the engine may incur damage if the indicator enters this section. On models with an LCD instrument panel display, the engine speed is shown in the form of a power curve on models up to and including 1990 – the maximum power occurs before maximum speed, and once maximum speed is reached, the entire curve flashes to indicate the danger zone.

Fuel gauge

Indicates the quantity of fuel remaining in the tank. When the needle enters the red band, the car should be refuelled. On models with an LCD instrument panel display, the fuel level is indicated by twelve segments – when the fuel level requires replenishment, the first segment alone remains lit up, and the amber warning zone surrounding the fuel pump motif flashes.

Temperature gauge

Indicates the engine coolant temperature. As the engine warms up, the needle should move from the blue (cold) scale into the black section. If the needle remains in the blue section, or enters the red (hot) section, a fault is indicated, and advice should be sought (refer to *Fault finding* on page 143).

Do not continue to run an engine which shows signs of overheating.

On models with an LCD instrument panel display, the temperature is indicated by six segments – the blue section indicates that the engine has not yet reached its normal working temperature, the yellow section indicates that the engine is at its normal temperature, and the red section indicates that the engine temperature is too high, and the engine should be switched off. With overheating, all segments light up, and in addition the red warning zone flashes.

Clock

An analogue quartz clock is fitted to some models. To adjust the time, depress the knurled knob in the centre of the clock and turn as required.

Oil pressure gauge

Fitted to some 'Mark 1' models (up to 1984), the oil pressure gauge can give early warning of trouble in the lubrication system. The scale ranges from 0 to 5 bars (0 to 71 lbf/in^2), and with an engine at normal temperature, the oil pressure must not drop below 2 bars (28.5 lbf/in^2) at driving speed.

On models with an LCD instrument display, oil pressure is indicated on a bar scale. When the oil pressure reaches the minimum allowable with the engine running, the red warning zone flashes. With the engine switched off, the warning zone lights up but does not flash – after the engine is started, the lights should go out. When the engine is hot, the lights may come on intermittently when idling, but they should go out when the engine speed is increased. If the warning zone flashes when driving, you must stop immediately and switch off the engine.

Voltmeter

Supplementing the alternator warning light on some models, the voltmeter indicates the state of charge of the battery, and also shows if the charging system is functioning correctly. When starting the engine, the battery voltage should not drop into the red zone, and during driving the indicator should stay in the black zone. Should this not be the case, and assuming that the alternator drivebelt is not broken, have the charging system and battery checked by an expert.

CONTROLS AND EQUIPMENT — 21

Warning lights

These lights warn the driver of a fault, or inform the driver that a particular device is in operation.

BRAKE FLUID LOW LEVEL/ HANDBRAKE 'ON' WARNING LIGHT

On some models, these two functions are on separate warning lights – the warning light with the letter 'P' lights up if the handbrake is applied with the ignition switched on, and the warning light with the exclamation mark lights up if the brake fluid level drops below the minimum mark. Where the two functions are combined, if the light comes on with the ignition switched on, check first that the handbrake is fully released. If the light does not then go out, stop immediately and check the brake fluid level (refer to 'Regular checks' on page 81).

Do not continue to drive the car if a brake fluid leak is suspected.

LOW OIL PRESSURE WARNING LIGHT

Warns that the engine oil pressure is low. If the light stays on for more than a few seconds after start-up, it's likely that the engine is worn, and advice should be sought. If the light comes on whilst driving, **switch off** the engine immediately and seek advice. When the engine is hot, the light may come on intermittently when idling, but it should go out when the engine speed is increased.

IGNITION WARNING LIGHT

Acts as a reminder that the ignition circuit is switched on if the engine isn't running, and acts as a no-charge warning light. If the light stays on after starting, or comes on whilst driving, the battery is not charging properly. The battery may therefore become discharged (go 'flat'), and the best course of action is to stop and seek advice.

MANUAL CHOKE WARNING LIGHT

Acts as a reminder that the choke is in operation. The light will go out when the choke control is pushed in.

HEADLIGHT MAIN BEAM WARNING LIGHT

Acts as a reminder that the headlight main beam is switched on. Will also light up when the headlamp flasher is operated.

DIRECTION INDICATOR WARNING LIGHT

Shows that the direction indicators are switched on. Rapid flashing of the warning light means that a bulb has blown.

TRAILER INDICATOR WARNING LIGHT

With the trailer or caravan electrics hooked up, this warning light should flash in unison with the car indicator warning light. Failure to do so indicates that a bulb has blown on the trailer lighting, or that there is a bad connection at the trailer plug and socket.

ENGINE CONTROL INDICATOR LIGHT

This indicates any fault in the electronic engine management system. It lights up when the engine is first switched on, and remains on during starting, but should go out when the engine is running. If the warning light comes on during driving, a fault has occurred, and the electronic system automatically switches to its emergency running programme. The car may be driven for a short distance with the warning light on, but prolonged driving may result in increased fuel consumption and lack of response from the engine. A Vauxhall dealer should be consulted as soon as possible. If the warning light comes on briefly and then goes out again, this is of no significance.

Check control warning system

Some models are fitted with a check control warning system which automatically checks fluid levels, brake pad lining thickness, and the main exterior lighting bulbs and wiring.

When the ignition is switched on, all the check control warning lights are illuminated for approximately 4 seconds. If all the systems are functioning correctly, the check control lights will all go out except for the brake light indicator, which will go out after the first application of the brake pedal. The systems are checked only when the ignition is switched on,

CONTROLS AND EQUIPMENT

since the movement of the car will cause the fluid levels to fluctuate. There are six systems on pre-1991 models, and seven systems on 1991-on models, and the warning lights will be indicated with the following faults.

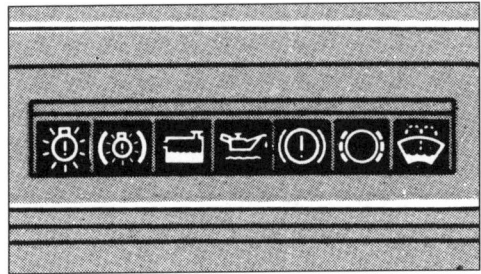

▲ Check control warning system

EXTERIOR LIGHT WARNING LIGHT

Indicates when a headlamp dipped beam bulb or a tail light bulb fails.

BRAKE STOP-LAMP WARNING LIGHT

Indicates when a brake stop-lamp bulb fails.

LOW COOLANT LEVEL WARNING LIGHT

Indicates that the engine coolant has reached the minimum acceptable level. The coolant level should be topped-up as soon as possible (refer to 'Regular checks' on page 80).

LOW ENGINE OIL LEVEL WARNING LIGHT

Indicates that the engine oil has reached the minimum acceptable level. Note that on pre-1991 models, the check is made only when the ignition is switched on, and will only be accurate if the car is on level ground. If, for example, the engine is stalled shortly after starting up from cold and then started again, an accurate reading will not be taken, since the oil will not have had a chance to drain back into the sump from the upper parts of the engine. On later models, the engine oil level is also monitored during driving. Should the light come on, the oil level should be topped-up as soon as possible (refer to 'Regular checks' on page 80). On later models, if the light flashes during driving, the engine should be switched off immediately and the oil level topped-up.

LOW BRAKE FLUID LEVEL WARNING LIGHT

Indicates that the brake fluid has reached the minimum acceptable level. The brake fluid level should be checked immediately (refer to 'Regular checks' on page 81).

Do not drive the car if a brake fluid leak is suspected.

BRAKE PAD WEAR WARNING LIGHT

Indicates that the front brake pads are worn to the stage where they require renewal. The pads should be renewed as soon as possible.

LOW WASHER FLUID LEVEL WARNING LIGHT

Indicates that the windscreen washer fluid has reached the minimum acceptable level.

Ignition switch/steering lock

The switch has four positions as follows:
- **B** Ignition off, steering locked
- **I** Ignition off, accessory circuits on, steering unlocked
- **II** Ignition on, and all electrical circuits on
- **III** Starter motor operates (release the key immediately the engine starts)

▲ Ignition switch/steering lock positions. Withdraw the ignition key in position **B** to lock the steering

To lock the steering, the ignition key must b withdrawn in position **B** – when the steering wheel is turned slightly, the lock will engage. To release the lock, insert the ignition key and

CONTROLS AND EQUIPMENT

turn it to position **I**, at the same time moving the steering wheel slightly from side to side.

Gearbox

MANUAL GEARBOX

Either a 4 or 5-speed gearbox may be fitted. The gear positions follow the usual 'H' pattern, with 5th gear to the right of the 3rd gear position, and reverse gear to the left of the 1st gear position.

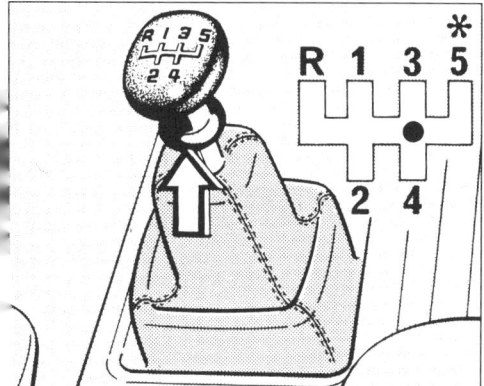

▲ *Gear lever positions with manual gearbox models. Lift ring (arrowed) to engage reverse gear*

To engage reverse gear, lift the ring below the gear knob before moving the gear lever into the reverse position. Only engage reverse gear with the vehicle stationary.

AUTOMATIC TRANSMISSION

There are six selector positions:

P – selects the **Park** position, which locks the automatic transmission. The engine can be started. Only select **P** after parking when the car is stationary, and always apply the handbrake.

R – selects **Reverse** gear. Only select **R** with the vehicle stationary.

N – selects **Neutral**. The engine can be started. No power is transmitted to the wheels.

D – selects **Drive**. The transmission will change gear automatically using 1st, 2nd and 3rd gears.

2 – allows the transmission to change automatically between 1st and 2nd gears only. Useful when driving up or down long hills, or when driving through a series of bends. Do not select **2** at speeds above approximately 65 mph.

1 – locks the transmission in 1st gear. It should only be used when driving up or down extremely steep gradients. Maximum engine braking is achieved, but if possible do not select **1** at speeds above approximately 34 mph. If **1** is selected above this speed, the transmission will remain in **2** until the vehicle is braked below 34 mph.

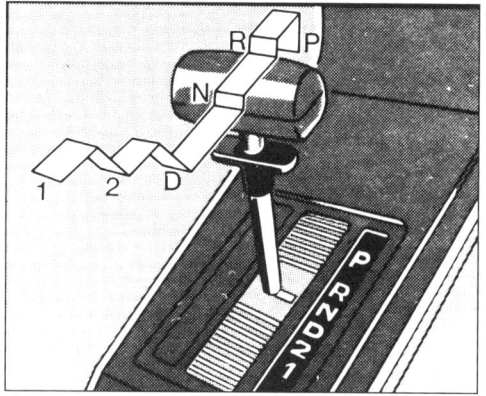

▲ *Automatic transmission selector positions*

The selector lever can be moved freely between positions **N** and **D**. To select any other position, the release collar below the selector handle must be pulled up. Pull the release collar up to the first stop to select **2**, and beyond the first stop to select **R**. Pull the collar

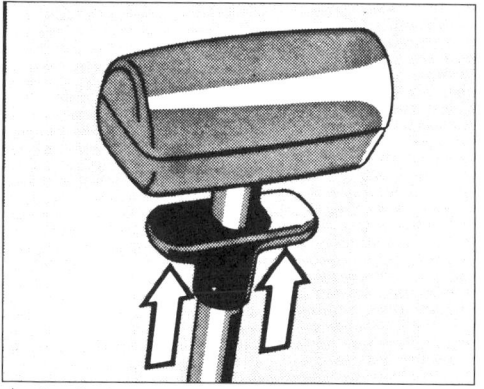

▲ *Lift collar (arrowed) when selecting **1**, **2**, **R** or **P** from a lower position*

VAUXHALL ASTRA & BELMONT

up to the second stop to select either **P** or **1**. Do not raise the collar when moving to a lower level position.

When the accelerator pedal is depressed fully to the floor (kickdown position) within certain speed ranges in 2nd and 3rd gears, the automatic transmission changes to a lower gear in order to give maximum acceleration.

Left-hand multi-function switch

Controls the direction indicators, headlight main and dipped beam, and headlight flash. Move the switch upwards to signal a right-hand turn, and downwards to signal a left-hand turn.

▲ Main and dipped beam selection

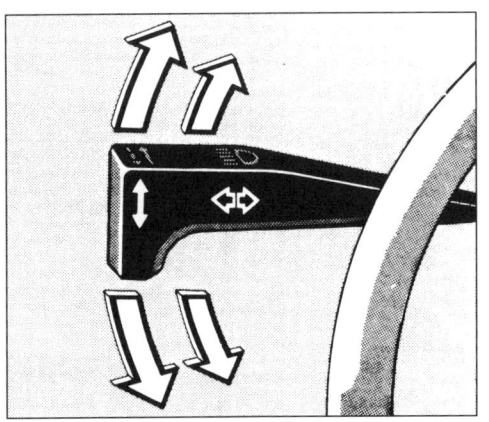

▲ Direction indicator selection

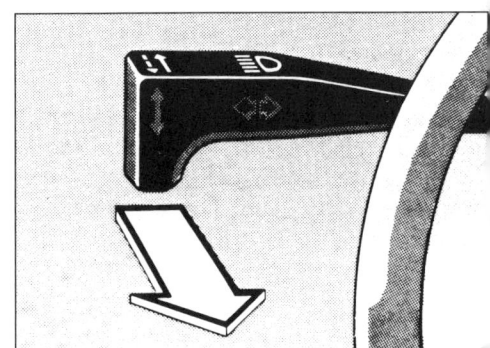

▲ Headlamp flash selection

The switch will automatically cancel when the steering wheel is returned to the straight-ahead position. When lane-changing or making minor steering movements, the cancel function will not operate, and for indication in these circumstances, the switch should be moved part-way to the first stop and held in this position by hand. The indicators will flash until such time as the switch is released.

The main beam position is selected by pushing the switch away from the steering wheel, and the dipped beam position by pulling the switch towards the steering wheel to the next position.

On early models fitted with auxiliary long-range headlights, these will operate automatically together with the main beams.

To operate the headlamp flash, pull the switch towards the steering wheel against the resistance.

Releasing the switch will turn the headlamp flash off. The headlamp flash may also be operated together with the direction indicators.

Exterior and interior lights switch

Controls the sidelights/tail lights/number plate lights and headlights, and also the interior light.

Turning the switch to the first position will operate the sidelights, and on models fitted with dim-dip lighting (1987 on) the headlights will glow dimly when the ignition is switched on. On later models, if the driver's front door is opened with the switch in either the sidelight main beam positions, when the ignition switch is in either the **B** or **I** positions, a warning buzzer will operate. Turning the switch to the

CONTROLS AND EQUIPMENT

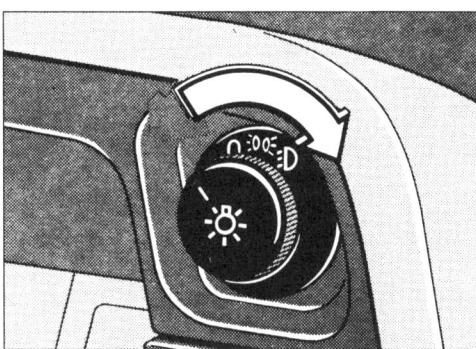

▲ Exterior lights switch

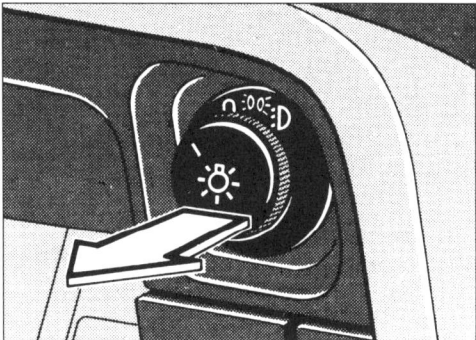

▲ Interior lights operation

second position will operate the headlights.
To operate the interior lights, the switch knob is pulled outwards. The interior lights come on automatically when the front doors are opened, and go out when the doors are closed. On models from 1986 onwards, a time delay is fitted so that the light remains on for a few seconds after closing the door.

Headlamp beam adjustment control

On certain models produced from 1991 onwards, a headlamp beam adjustment control is fitted to the facia, just to the right of the steering wheel.
The control has four settings as follows.

HATCHBACK AND SALOON MODELS

0 Driver's seat occupied
1 All seats occupied
2 All seats occupied and load in luggage compartment
3 Driver's seat occupied and load in luggage compartment

ESTATE MODELS

0 Seats occupied
1 Not used
2 Seats occupied and load compartment half-loaded
3 Seats occupied and load compartment fully-loaded

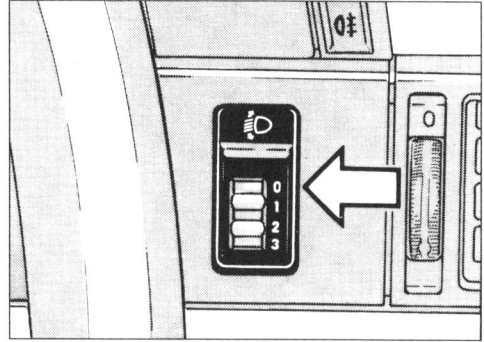

▲ Headlamp beam adjustment control

Right-hand multi-function switch

Controls the windscreen wipers and washers. Where a headlight wash/wipe system is fitted, this will operate whenever the screen washer is operated with the sidelights or headlights switched on. On models from 1983 onwards, the switch also operates the rear window wash/wipe system – on models produced before this date, the rear wash/wipe system is operated by a switch on the facia.

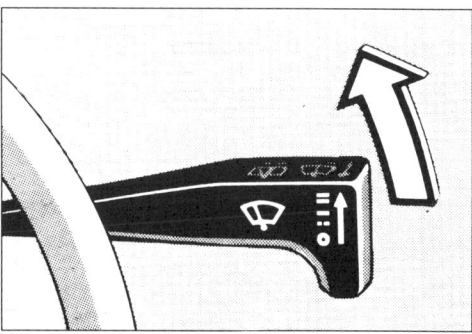

▲ Windscreen wiper operation

VAUXHALL ASTRA & BELMONT

26 CONTROLS AND EQUIPMENT

To operate the windscreen wipers, move the switch upwards from its off position.

The first position will operate the wipers intermittently, pausing for several seconds between each wipe. The second position will operate the wipers continuously at a slow speed, and the third position will operate them continuously at a fast speed. To operate the washers, pull the switch towards the steering wheel, and release the switch to stop the washers.

The wipers operate automatically for several wipes while the washers are operating, and on models so equipped, the headlight wipers and washers will operate at the same time if the sidelights or headlights are switched on.

On 1983-on models fitted with a rear wiper, push the switch away from the steering wheel to the first stop to operate the rear wiper intermittently, and to the second stop to cause the washers to operate automatically as the rear wiper is in operation.

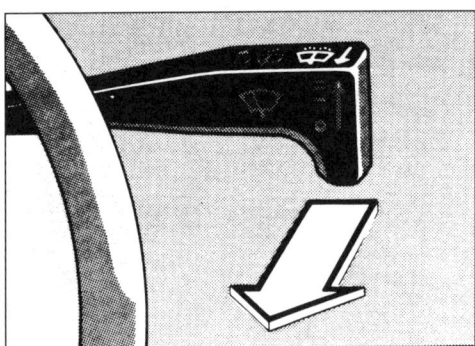
▲ Windscreen washer operation

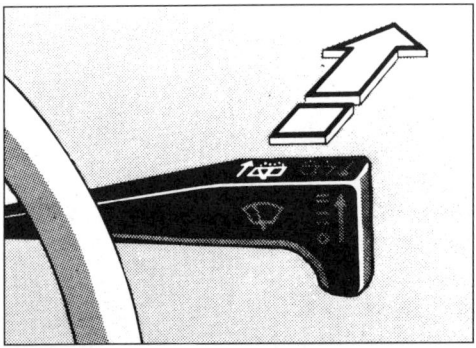
▲ Rear wipe/wash operation for models from 1983-on

On earlier models, raise the facia switch to its first position to operate the rear wiper, then raise it against the spring pressure to the second position to operate the washers.

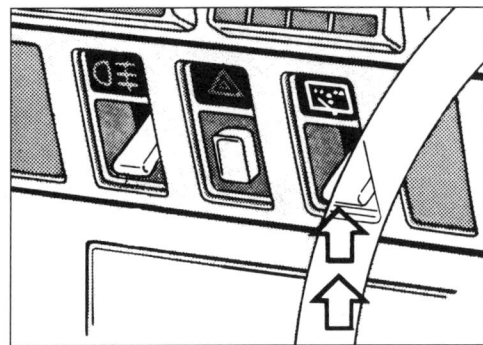
▲ Rear wipe/wash operation for models before 1983

Heating and ventilation controls

The rotary knob located below ('Mark 1' models, up to October 1984) or next to ('Mark 2' and Belmont models, from October 1984) the heater control sliders, operates the heater blower.

▲ Heater blower control knob

Positioned fully anti-clockwise, the blower is switched off – turning it clockwise operates the blower at two (up to 1982 models) or three (1983-on models) speeds from slow to fast.

The left-hand heater slide lever controls the air temperature. Maximum heat is obtained

VAUXHALL ASTRA & BELMONT

CONTROLS AND EQUIPMENT

with the lever fully upwards. Since the heater obtains its heat from the engine coolant, the engine must have been running for several minutes before any output can be expected. The right-hand heater slide lever(s) control the air distribution. On 'Mark 1' models there is only one lever – later models have two levers.

On models up to 1982, ventilation is closed off with the right-hand lever in the lower position, directed to the foot area in the centre position, and directed to the head and windscreen areas in the upper position. The air direction is fully variable between the foot and head areas (between the two arrow positions).

On models from 1982 to 1984, ventilation is directed to the head area with the right-hand lever in the lower position, directed to the foot area in the centre position, and directed to the windscreen area in the upper position. Between the centre and upper positions the air direction is fully variable between the windscreen and foot areas.

▲ Heater air temperature and distribution control levers on later ('Mark 2') models

▲ Central air vents

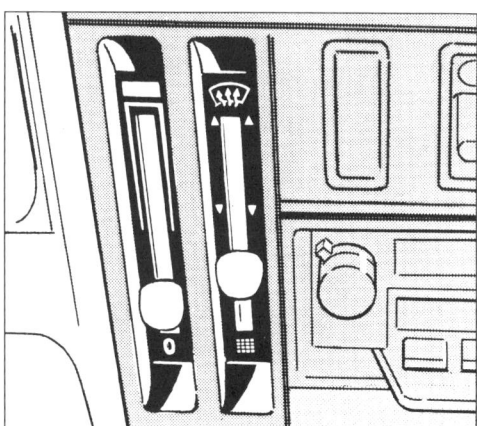
▲ Heater air temperature and distribution control levers on early models

With the later ('Mark 2') two-lever air distribution system, the upper lever directs air to the screen (upwards) or head space (downwards). The lower lever directs air to the head area (upwards) or foot area (downwards).

Fresh unheated air can be directed into the vehicle through the central vents.

The air supply is regulated by means of the control wheel, and its direction can be controlled by tilting the vents and adjusting the

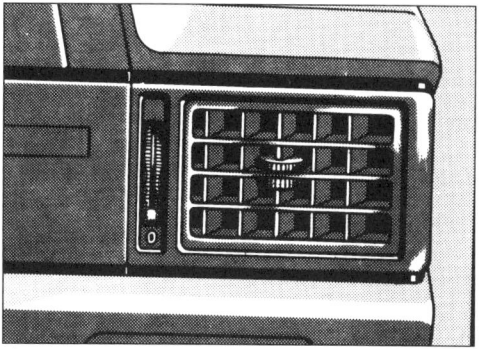
▲ Side air vent

fins as required. The air volume is controlled by the heater blower. The side vents operate in a similar manner.

VAUXHALL ASTRA & BELMONT

CONTROLS AND EQUIPMENT

The direction of air to the front door windows for defrosting or demisting is controlled by moving the right-hand distribution lever(s) upwards. Hot or cold air may be supplied depending on the position of the left-hand lever.

▲ Front door window demister vent

▲ Hazard flasher switch on 'Mark 1' models

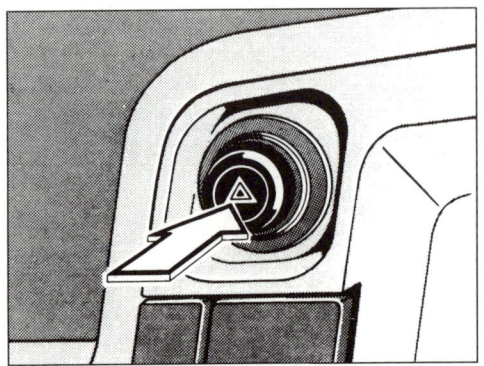
▲ Hazard flasher switch on 'Mark 2' and Belmont models

Hazard flasher switch

Operates all direction indicators in unison, regardless of the position of the ignition switch. Depress the switch once to switch on the hazard lights, and depress the switch again to switch them off.

Rear foglight switch

Operates the rear foglights. On most models the sidelights must be switched on for the rear foglights to operate, and on some models the rear foglights will go out when the headlamp main beam is switched on.

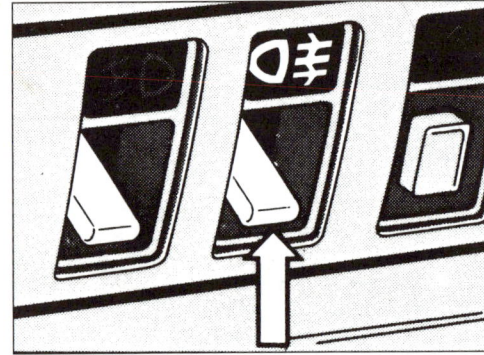
▲ Rear foglight switch on 'Mark 1' models

▲ Rear foglight switch on 'Mark 2' and Belmont models

Front foglight switch

Operates the front foglights, provided that the sidelights and ignition are switched on. On some models the front foglights are extinguished when the headlamp main beam is switched on.

VAUXHALL ASTRA & BELMONT

CONTROLS AND EQUIPMENT

Heated rear window switch

Operates the heated rear window, provided that the ignition is switched on.

▲ *Heated rear window switch operation*

The switch is incorporated in the heater blower rotary knob – pull the knob out to operate the heated rear window. To avoid unnecessary electrical loads, it's recommended that the device is switched off as soon as demisting is complete.

Instrument illumination control

Only fitted to 'Mark 2' and Belmont models, this controls the brightness of the instrument panel illumination when the exterior lights are switched on. The control is located to the left of the instrument panel below the hazard lighting switch, and is in the form of a horizontal wheel. On 'Mark 1' models, the illumination is not adjustable.

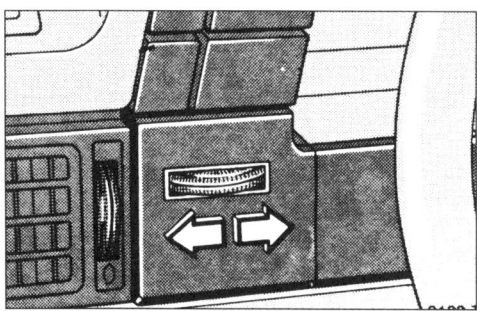

▲ *Instrument illumination control*

Electric door mirror switch

On models fitted with electric door mirrors, the controlling switch is located on the centre console.

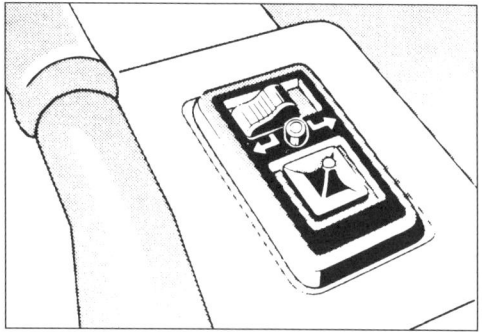

▲ *Electric door mirror switch*

The rocker switch determines which mirror is to be adjusted – pushing the switch to the left will control the left-hand mirror, and *vice-versa*. Adjust the mirror by moving the four-way control knob.

Where the mirrors are heated, depressing the button on the switch will heat both mirrors for approximately 15 minutes. The green warning light will indicate that the mirrors are being heated.

Bonnet release lever

Mounted beneath the left-hand side of the instrument panel on 'Mark 1' models, or beneath the right-hand side of the instrument panel on 'Mark 2' and Belmont models, pull the lever to release the bonnet.

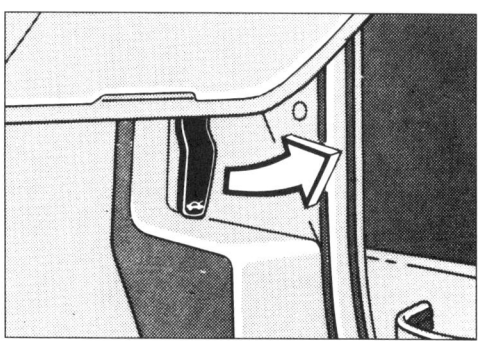

▲ *Bonnet release lever under the right-hand side of the instrument panel*

CONTROLS AND EQUIPMENT

Lift the bonnet safety catch located under and slightly to the left of centre of the bonnet (viewed from in front of the car), then open the bonnet and support it using the stay located in the special slot.

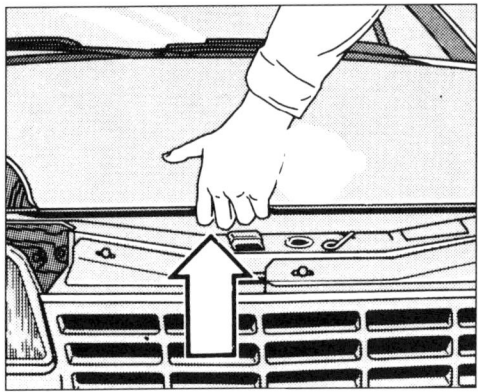
▲ *Safety catch under the centre of the bonnet*

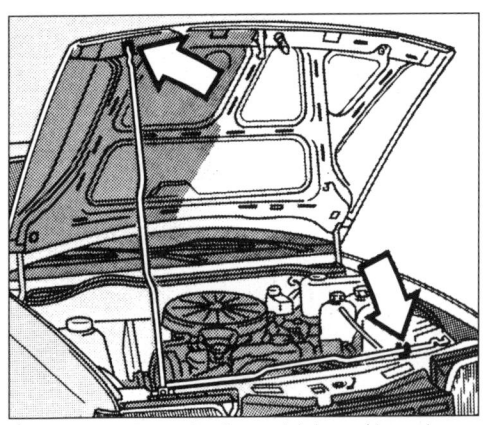
▲ *Support stay located in the special slot and in retainer*

On 'Mark 1' models, the stay is located on the right-hand side of the engine compartment (as viewed from the front of the car), but on later models it is located on the opposite side.

When closing the bonnet, make sure that the release lever inside the car is fully returned to its rest position. Check that the stay is firmly in its retainer, then lower the bonnet until it is approximately six inches above the lock, and let it drop the final distance. Finally press on the bonnet to confirm that it is firmly locked.

Tiltable steering column

'Mark 2' and Belmont models may be fitted with a tiltable steering column.

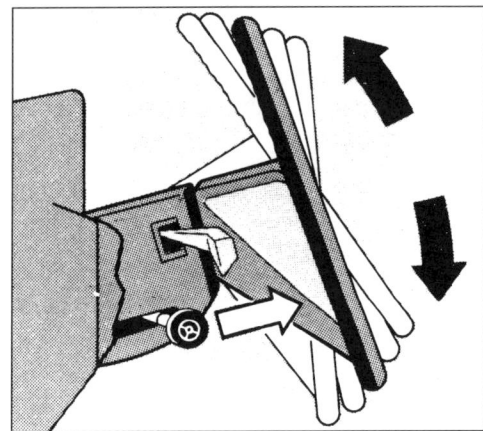
▲ *Tiltable steering column adjustment*

The upper steering column, together with the steering wheel, may be adjusted up or down for the most comfortable position. Move the locking lever located on the left-hand side of the column upwards towards the steering wheel, then adjust the steering wheel to one of the five positions, and lock it by pushing the lever down again.

INTERIOR EQUIPMENT

Cigarette lighter

With the ignition switched on, press the lighter in then release it, and wait until it pops out ready for use.

Electric windows

On models fitted with electric windows, the operating switches are located between the front seats on the centre console.

Depending on the model, either two or four switches may be fitted, for operation of the front door windows only, or both front and rear door windows. The switches are of the rocker type – depressing the rear edge opens the windows, and depressing the front edge closes the windows.

VAUXHALL ASTRA & BELMONT

CONTROLS AND EQUIPMENT 31

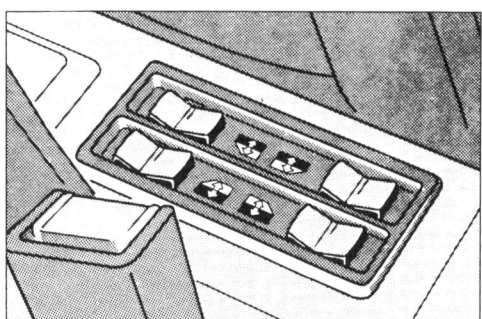

▲ Electric window switches

Sunroof

The sunroof is either of the sliding or hinged (removable) type. On 'Mark 1' models, the sliding type is operated by a hinged handle which is used to lock the sunroof in the desired position, but later models have a cranked handle which, when turned, lifts the rear edge of the sunroof or moves it to the desired position. An adjustable sunshade is located beneath the sunroof on these later models, but this should not be used when the sunroof is completely open.

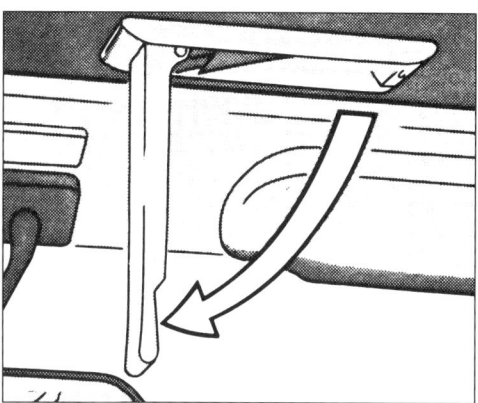

▲ Sliding sunroof operating handle on 'Mark 1' models

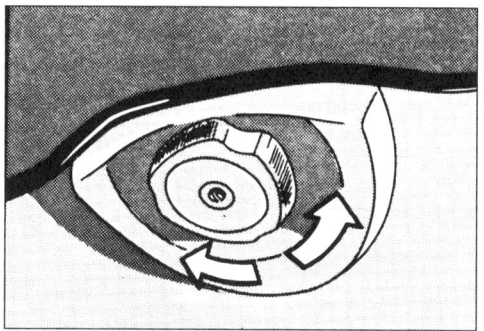

▲ Hinged sunroof operating control knob

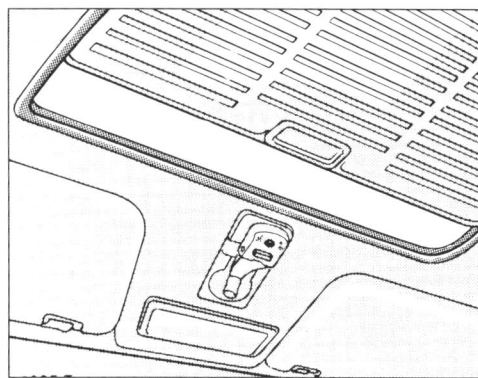

▲ Adjustable sunshade

To open the 'Mark 1' sliding roof, pull down the handle and move the sunroof rearwards to the desired position. Lock the sunroof by pushing the lever up into its recess.

To raise or lower the hinged type sunroof, turn the control knob as required.

To remove the sunroof, close it and turn the screw in the centre of the control knob clockwise using a coin or similar tool.

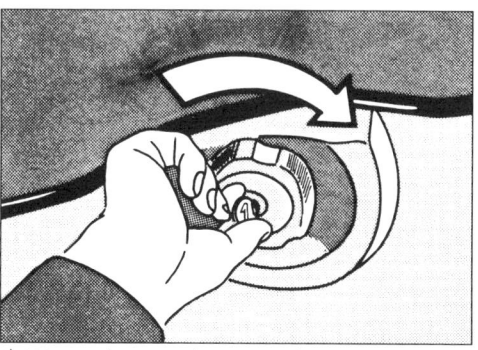

▲ Using a coin to turn the screw in the control knob

Raise the roof slightly and disengage the safety catch by pressing the release lever, then pivot it and remove it from its retainers.

A special storage pouch is provided for the roof panel, on the back of the rear seat.

VAUXHALL ASTRA & BELMONT

CONTROLS AND EQUIPMENT

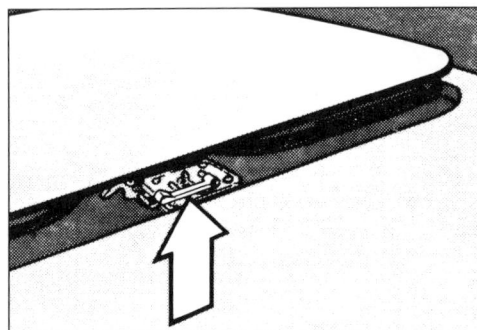

▲ Safety catch on the hinged sunroof

▲ Removing the hinged sunroof

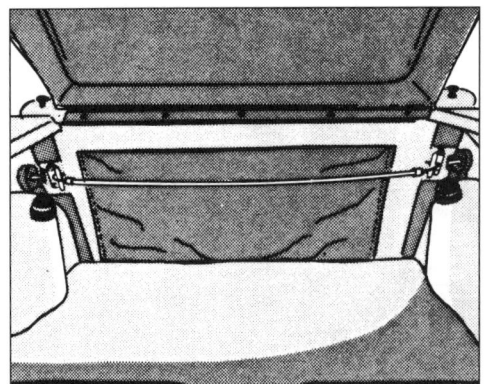

▲ Storage pouch for sunroof on back of seat

engage the safety catch and close the roof by using the control knob.

To open the later sliding roof, pull the cranked handle from its recess then unlock it by pressing the button.

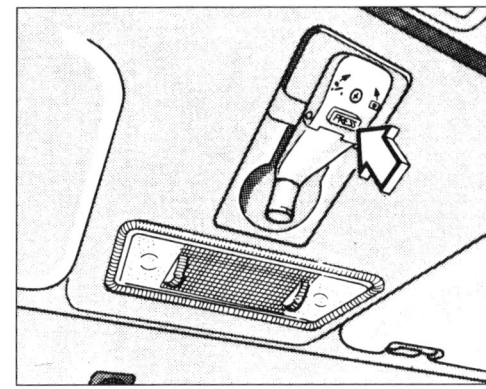

▲ Unlocking button on the cranked handle on later sliding sunroof

Turn the handle anti-clockwise to move the sunroof rearwards to the desired position, then press the handle back into its recess to lock it. If it is only required to raise the rear edge of the sunroof for ventilation (ie without moving it rearwards), turn the handle clockwise then press it back into its recess.

Dipping rear view mirror

Pull back the lever under the mirror to reduce the glare from the lights of following vehicles.

Seat belts

The operation of the seat belts is self-explanatory, but certain points require additional explanation.

FRONT SEAT BELT HEIGHT ADJUSTMENT

On certain models, the height of the upper seat belt mounting can be adjusted. On models produced up to 1988, this is achieved by depressing the button and moving the mounting to the required setting.

On models from 1989, it is necessary to tilt the attachment on the pillar and move it to the required setting.

Stow the panel with the front hinge tongues upwards and the curved surface towards the back of the seat. When refitting the roof panel, turn the control knob to the open position. Fit the front hinge tongues into their retainers,

VAUXHALL ASTRA & BELMONT

CONTROLS AND EQUIPMENT   33

▲ Front seat belt height adjustment on models up to and including 1988

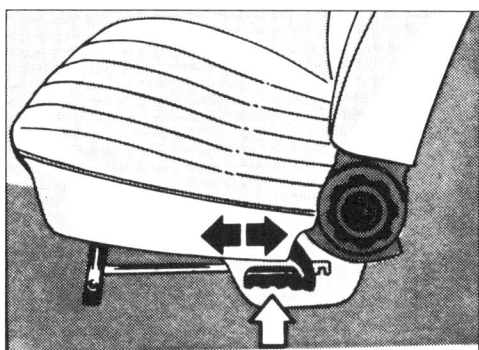

▲ Front seat position adjustment on 'Mark 1' models

▲ Front seat belt height adjustment on models from 1989

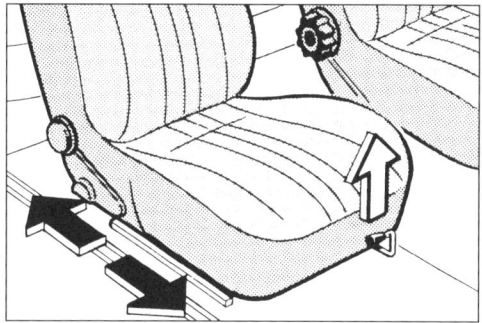

▲ Front seat position adjustment on 'Mark 2' and Belmont models

REAR SEAT BELTS

Where fitted, the centre rear seat belt is of the lap type (non-inertia), and must be manually adjusted by the wearer. To adjust its length, depress the black button above the lock tongue and pull the webbing through as required.

Front seat adjustment

SEAT SLIDING CONTROL

To move the front seat backwards or forwards, pull up the lever located on the base of the seat back ('Mark 1' models) or on the front of the seat (later models), then slide the seat to the required position and release the lever to lock it.

SEAT BACK RELEASE (THREE-DOOR ASTRA MODELS)

Pull up the lever on the outside of the front seat backrest, and swivel the backrest forwards to allow rear seat passengers to enter.

The backrest will click into place when it is swivelled back to its normal position.

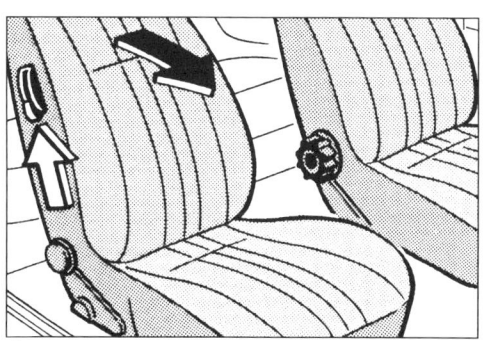

▲ Front seat backrest release

VAUXHALL ASTRA & BELMONT

CONTROLS AND EQUIPMENT

SEAT RECLINING CONTROL

Turn the wheel on the inside rear base of the seat to alter the angle of the seat back. Do not lean back on the seat while making the adjustment.

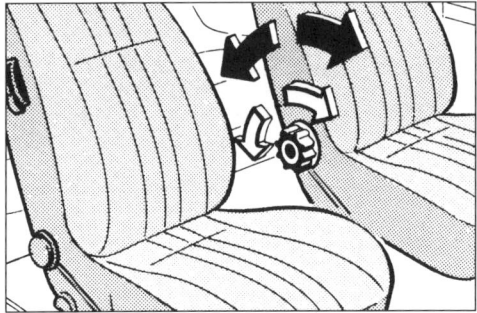

▲ *Front seat reclining control adjustment*

FRONT SEAT HEIGHT ADJUSTMENT

On some models the height of the front seat is adjustable by turning the handle located on the front of the seat.

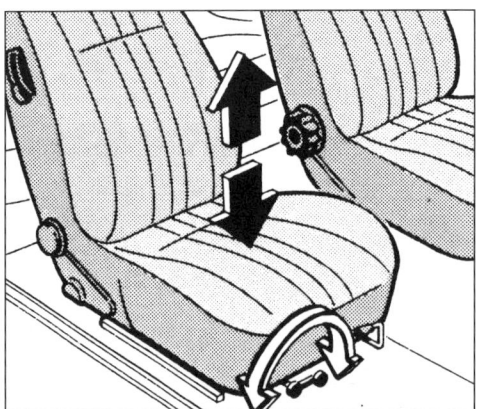

▲ *Front seat height adjustment*

Adjustable head restraints

The head restraints can be adjusted for height – the upper edge of the restraint should be approximately at ear height.

On models up to and including 1982, push the top of the restraint rearwards to unlock it, then position it as required and lock it by pulling the top forwards.

On models manufactured in 1983 and 1984, simply move the restraint to the required position – it will be locked in the vertical position automatically. To remove the restraint, first prise the retaining spring from the left-hand guide using a screwdriver, then lift the restraint from the back of the seat.

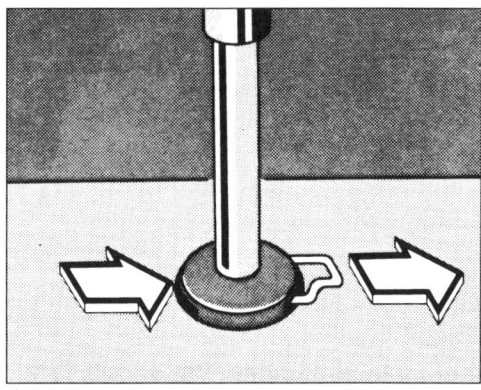

▲ *Pull the retaining spring out to remove the head restraint on models manufactured in 1983 and 1984*

When refitting, make sure that the spring is inserted with its bent side pointing towards the rear.

On 'Mark 2' and Belmont models, adjustment is by moving the restraint to the required position, although in some cases it may help to keep the retaining springs depressed while making the adjustment. To remove the restraint, release the catch retaining springs by pushing them to the rear.

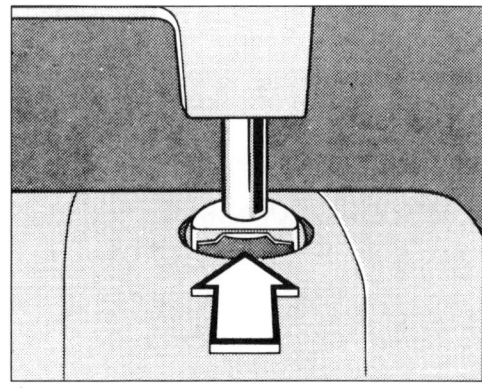

▲ *Head restraint retaining spring on 'Mark 2' and Belmont models*

VAUXHALL ASTRA & BELMONT

CONTROLS AND EQUIPMENT

Folding rear seats

HATCHBACK MODELS

To fold down the rear seat on Hatchback models, first unhook the rear shelf cords from the tailgate. Place the central seat belts between the backrest and the cushion, and secure the side seat belts with the special clips.

▲ Secure the side seat belts in the special clips

Release the catches on the top of the backrest, and fold the backrest together with the rear shelf onto the cushion.

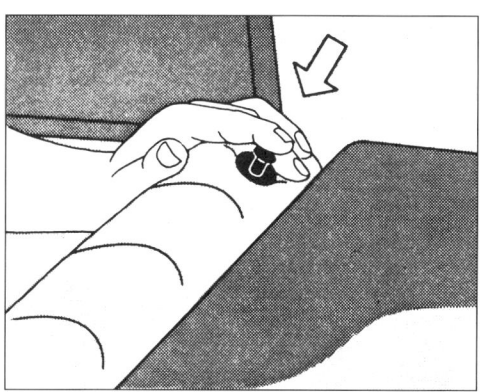

▲ Depress the buttons to fold the rear seat backrest forwards

If long objects are being carried, leave the rear seat backrest folded down, then move the front passenger seat fully forwards and tilt its backrest fully to the rear.

For maximum load-carrying capacity, first pull the seat belts through to the rear. On 'Mark 1' models, fold the backrest forwards onto the seat cushion, then pull up the cushion by means of the strap so that the backrest and cushion are positioned vertically behind the front seats. On 'Mark 2' models, pull up the cushion using the strap provided, and position it vertically behind the front seats, then fold the backrest flat.

▲ Rear seat positions for 'Mark 1' Hatchback models

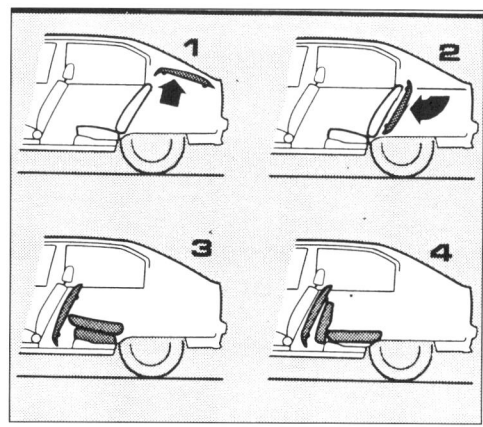

▲ Rear seat positions for 'Mark 2' Hatchback models

When repositioning the rear seat, make sure that it locks audibly in the catches.

ESTATE MODELS

To fold down the rear seat on Estate models, first locate the seat belts in the special clips.

CONTROLS AND EQUIPMENT

▲ *Rear seat belt securing clip on Estate models*

For carrying long objects, release the catches on top of the backrest and fold down onto the cushion, then adjust the front passenger seat fully forwards and tilt its backrest fully to the rear.

For carrying other objects in the luggage compartment, fold the seat cushion forwards by pulling it up with the strap provided, then release the catches on the top of the backrest and fold down flat onto the cushion.

When re-positioning the rear seat, make sure that it locks audibly in the catches.

SALOON (BELMONT) MODELS

To fold down the rear seat on Belmont models, release the catches on top of the backrest and fold the seat forwards.

The seat is in two sections, so either section may be folded down independently of the other, or both sections may be folded down together. Make sure that the seats lock audibly in the catches when re-positioning them.

CONVERTIBLE MODELS

To fold down the rear seat on Convertible models, if the roof is open, first release the press fasteners and Velcro strip of the roof cover from the rear seat backrest. The release buttons for the rear seat backrests are located inside the boot on the upper right-hand side – pull the buttons as necessary to release the appropriate backrest, and fold forwards onto the cushion.

When re-positioning the rear seat, make sure that it locks audibly in the catches.

▲ *Rear seat backrest release buttons on Convertible models*

EXTERIOR EQUIPMENT

Door locks

All doors may be locked from the inside by depressing the locking knob. If the knob is depressed while the door is open, all doors except the driver's door will lock when they are closed. The driver's door knob will spring up to its unlocked position when the door is closed, to prevent the driver from being locked out. To lock the driver's door, first close the door then turn the key in the lock.

The rear door locks incorporate childproof catches, operated by pushing up the small lever on the rear edge of the door below the lock. The rear doors cannot then be opened from

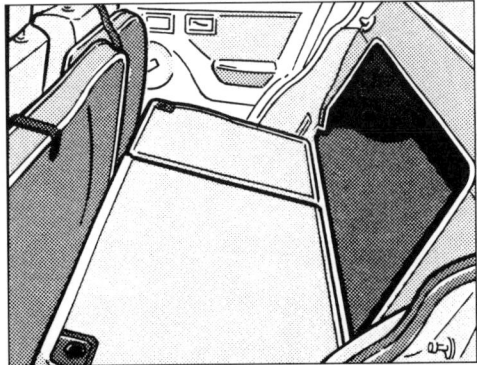

▲ *Rear seat folded down on Belmont models*

VAUXHALL ASTRA & BELMONT

CONTROLS AND EQUIPMENT

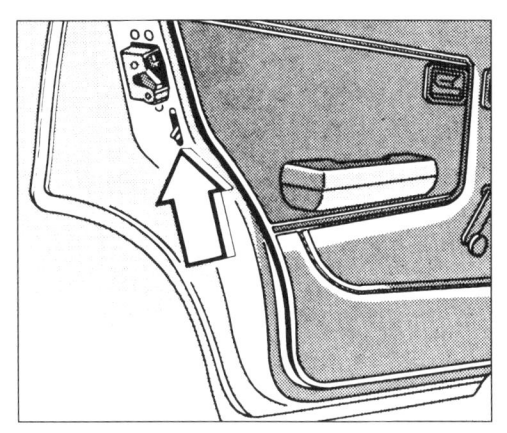

▲ *Childproof catch on the rear doors*

inside the car, although they may be opened from outside the car, provided that the locking knob has not been depressed.

Central door locking system

On models fitted with central locking, all doors are simultaneously locked when the driver's door is locked by turning the key in the lock or by depressing the locking knob inside the car. On Hatchback models, this includes the tailgate, but it is possible to unlock and lock the tailgate separately from the rest of the car. The key cannot be removed from the lock unless the tailgate is shut, so that the central locking system remains operative.

On Belmont models, if the boot is to be locked using the central locking system, the key slot in the boot lock must be horizontal. If the boot has been locked by the key, the slot will be vertical, and in this position the boot remains always locked. If the boot has been unlocked by key after having locked it with the central locking system, the key cannot be removed until it has been returned to the horizontal position. This system also applies to some later Hatchback models.

If the central locking motors are overloaded, the power supply to them will automatically be cut off for approximately 30 seconds.

Level control system

Certain models are fitted with a level control system, which helps maintain a constant vehicle level with various loads. The system should be set up as follows.

Before loading the vehicle, check the pressure at the level control valve (near the tailgate or boot opening) using a tyre pressure gauge. Adjust if necessary to 0.8 bars (11 lbf/in^2), using a footpump, bicycle pump, or garage air line.

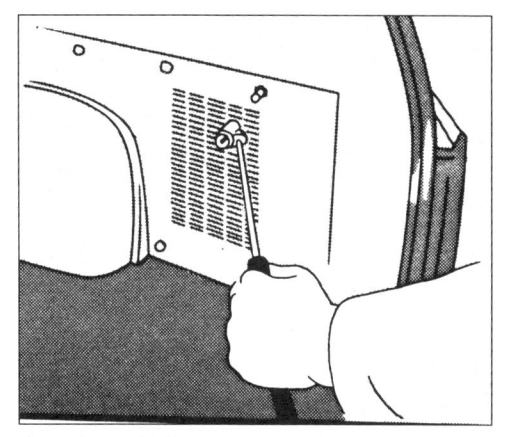

▲ *Level control inflation point – 'Mark 1' models*

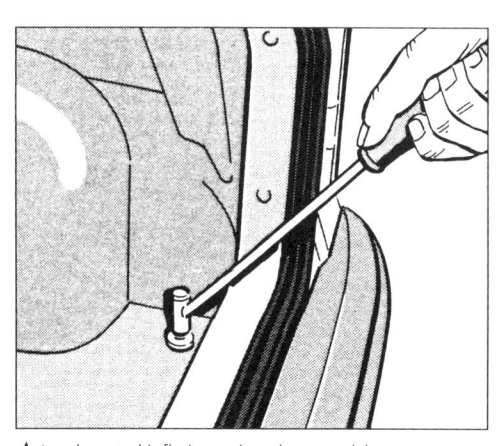

▲ *Level control inflation point – later models*

With the car on level ground, measure the distance between the centre of the rear bumper and the ground – this measurement is the car 'ride height'.

Load the vehicle, then pump up the system (again using a footpump/bicycle pump or garage air line) until the previously-measured ride height is restored. However, **do not**

VAUXHALL ASTRA & BELMONT

CONTROLS AND EQUIPMENT

exceed a pressure of 5 bars (71 lbf/in²) – check the pressure as you go.

When the load is removed, let air out of the control valve until the correct height is restored, but do not reduce the pressure below 0.8 bars (11 lbf/in²).

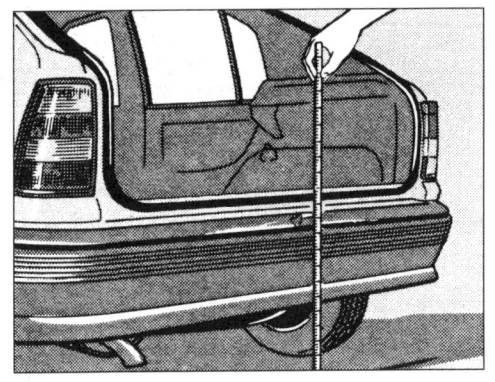

▲ *Measuring the rear ride height*

Folding roof (Convertible models)

To open the roof, first clear any items off of the rear parcel shelf. Push up the safety catches on the locks attaching the roof to the windscreen surround, then pull down the two handles to release the hooks.

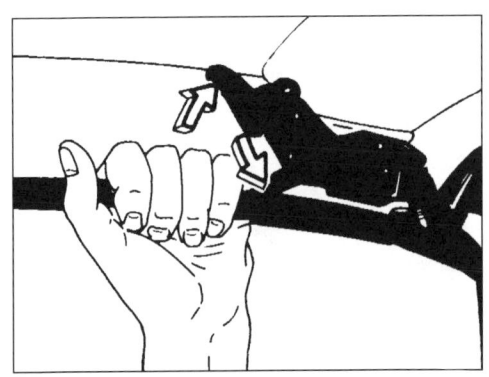

▲ *Releasing the folding roof catches from the windscreen surround*

Grasp the roof by one of the handles and fold it back – it will automatically fold into the correct position. Push down the two side sections of the roof until the levers are felt to lock.

▲ *Push down and lock the side sections*

To fit the cover, locate it loosely into position, then press in the two clips. Press the Velcro strip against the backrests, and press on the fasteners. Insert the rear hooks into the loops.

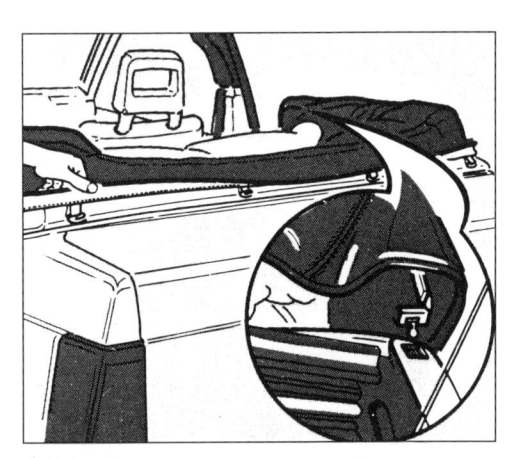

▲ *Fitting the cover*

To close the roof, remove the top cover and stow it in the luggage compartment. Disengage the locking levers by pushing them forwards, then grasp one of the handles and raise the roof to its normal position. Working on one side at a time, release the catch and attach the hook to the windscreen surround, then push up the handle to lock it in position. Note that the catches must always be locked when the car is being driven.

VAUXHALL ASTRA & BELMONT

ACCIDENTS AND EMERGENCIES

In the event of an accident, the first priority is safety. This may seem obvious, but in the heat of the moment, it's very easy to overlook certain points which may worsen the situation, or even cause another accident. The course of action to be taken will vary depending on how serious the accident is, and whether anyone is injured, but always try to think clearly, and don't panic.

We don't suggest that you consult this Section at the scene of an accident, but hopefully the following advice will help you to be better prepared to deal with the situation should you be unfortunate enough to appear on the scene of an accident or become involved in one yourself.

HOW TO COPE WITH AN ACCIDENT

Deal with any possible further danger

Further collisions and fire are the dangers in a road accident.

- **If possible warn other traffic**

    Where possible, switch on the car's hazard warning flashers.

    If a warning triangle is carried, position it a reasonable distance away from the scene of the accident, to give approaching drivers sufficient warning to enable them to slow down. Decide from which direction the approaching traffic will have least warning, and position the triangle accordingly.

    If possible, send someone to warn approaching traffic of the danger which exists, and to encourage the traffic to slow down.

- **Switch off the ignition and impose a 'No Smoking' ban**

    This will reduce the possibility of fire, should there be a petrol leak.

Call for assistance

Send someone to call the emergency services (dial 999), and make sure that all the necessary information is provided to the operator. Give the exact location of the accident, and the number of vehicles and if applicable the number of casualties involved. Refer to the *'Motorway breakdowns'* Section on page 48 for details of how to call for assistance on a motorway.

- **Call an ambulance**

    If anyone is seriously injured or trapped in a vehicle.

- **Call the fire brigade**

    If anyone is trapped in a vehicle, or if you think that there is a risk of fire.

- **Call the police**

    If any of the above conditions apply, or if the accident is a hazard to other traffic. In most cases, the accident must be reported to the police within 24 hours even if none of the above conditions apply (refer to *'Requirements of the law in the event of an accident'* on page 42).

Administer first aid

Refer to *'First aid'* on page 40.

Provide your details

If you are involved in the accident as a driver or car owner, provide your personal and vehicle details to anyone having reasonable grounds to ask for them. Also inform the police of the details of the accident as soon as possible, if not already done.

VAUXHALL ASTRA & BELMONT

ACCIDENTS AND EMERGENCIES

FIRST AID

Always carry a first aid kit
If possible, learn first aid by attending a suitable course – contact the St John Ambulance Association or Brigade, St Andrew's Ambulance Association or the British Red Cross Society in your local area for details. These organisations will also be able to provide further training in heart massage and mouth-to-mouth resuscitation which could enable you to save someone's life in an emergency.

Before proceeding with any kind of first aid treatment, deal with any possible further danger, and call for assistance, as described previously in this Chapter.

To help remember the sequence of action which should be followed when dealing with a seriously injured casualty, use the **ABC** of emergency first aid.

Casualties remaining in vehicles
Any injured people remaining in vehicles should not be moved unless there is a risk of further danger (such as from fire, further collisions etc).

Casualties outside the vehicles

Recovery position
To prevent the possibility of an unconscious but breathing casualty choking or suffocating, he/she must be placed in the recovery position.

▲ *The recovery position*

Lie the casualty on his/her side. Bend the leg and position the arms as shown to support the head.

First aid for other injuries
Always treat injuries in the following order of priority:
- Ensure that the casualty has a clear airway. If not treat as described opposite.
- Put any breathing, unconscious casualty into the recovery position as described previously.
- Treat any severe bleeding by applying direct pressure to the wound. Maintain the pressure for at least 10 minutes and apply a suitable clean dressing. If the wound continues to bleed, apply further pressure and dressings **over** that already in place. If bleeding from a limb, and as long as the limb is not broken, lift the affected limb to reduce the bleeding.
- Broken limbs should not be moved, unless the casualty has to be moved in order to avoid further injury. Support the affected limb(s) by placing blankets, bags, etc, alongside.
- Burns should be treated with plenty of cold water as soon as possible. Don't attempt to remove any clothing, but cover with a clean dressing.

Reassurance
The casualty may be in shock, but prompt treatment will minimise this. Reassure the casualty confidently, avoid unnecessary movement, and keep him/her comfortable and warm. Make sure that the casualty is not left alone.

Give the casualty NOTHING to eat, drink or smoke.

VAUXHALL ASTRA & BELMONT

ACCIDENTS AND EMERGENCIES

A Airway
● **Check for breathing**

Check that the airway is clear. If the casualty is breathing noisily or appears not to be breathing at all, remove any obvious obstruction in the mouth and tilt the head back as far as possible. Breathing may then start.

B Breathing
● **If breathing has stopped**

If after clearing the airway the casualty still appears not to be breathing, look to see if the chest or abdomen is moving. Place your ear close to the casualty's mouth to listen and feel for breathing, and look to see if there's any movement of the chest. If you can't detect anything, you must start breathing for the casualty (mouth-to-mouth resuscitation).

To do this, pinch the casualty's nostrils firmly, keep his/her chin raised, and seal your lips around the casualty's mouth. Breathe out through your mouth into the casualty until the chest rises, then remove your mouth and allow the casualty's chest to fall. Give one further breath.

If giving mouth-to-mouth resuscitation to a child, remember that an adult's lungs are significantly larger than those of a child. Care must be taken not to overinflate a child's lungs, as this may cause injury.

● **Check for a pulse at the base of the neck**

If the heart is beating, continue to breathe into the casualty at a rate of one breath every 5 seconds, until he/she is able to breathe unaided, then place the casualty in the recovery position (refer to *'Recovery position'* opposite).

C Circulation
● **If the heart has stopped**

If there is no pulse at the neck, then external chest compression (heart massage) must be started.

To do this, if the casualty is still in the vehicle, he/she must be removed and laid on his/her back on the ground. Kneel down on one side of the casualty. Feel for the lower half of the casualty's breastbone and place the heel of one of your hands on this part of the bone, keeping your fingers off the casualty's chest. Cover this hand with your other hand, interlocking your fingers. Keeping your arms straight and vertical, press the breastbone down about 4 to 5 centimetres (1½ to 2 inches). The pressure should be smooth but not jerky. Do this 15 times, at the rate of just over once a second. It may help you to count aloud as you press.

Follow this by 2 breaths as described under *'Breathing'*.

Repeat the cycle of 2 breaths followed by 15 compressions, twice more, ending with 2 breaths, then re-check the pulse.

● **If there is still no pulse**

Repeat the cycle 9 times and then check the pulse. Continue this procedure until there is a pulse or until professional help arrives.

ACCIDENTS AND EMERGENCIES

REQUIREMENTS OF THE LAW IN THE EVENT OF AN ACCIDENT

The following is taken from the Road Traffic Act of 1988.

If you are involved in an accident – which causes damage or injury to any other person, or another vehicle, or any animal (horse, cattle, ass, mule, sheep, pig, goat or dog) not in your vehicle, or roadside property:

You must
- stop;
- give your own and the vehicle owner's name and address and the registration mark of the vehicle to anyone having reasonable grounds for requiring them;
- if you do not give your name and address to any such person at the time, report the accident to the police as soon as reasonably practicable, and in any case within 24 hours;
- if anyone is injured and you do not produce your certificate of insurance at the time to the police or to anyone who has with reasonable grounds required its production, report the accident to the police as soon as possible, and in any case within 24 hours, and either produce your certificate of insurance to the police when reporting the accident or ensure that it is produced within seven days thereafter at any police station you select.

ESSENTIAL DETAILS TO RECORD AFTER AN ACCIDENT

If you're involved in an accident, note down the following details which will help you to complete the accident report form for your insurance company, and will help you if the police become involved.
- The name and address of the other driver and those of the vehicle owner, if different.
- The name(s) and address(es) of any witness(es) (independent witnesses are particularly important).
- A description of any injury to yourself or others.
- Details of any damage caused to the vehicles involved or other property.
- The name and address of the other driver's insurance company and, if possible, the number of his/her certificate of insurance.
- The registration number of the other vehicle (check this against the tax disc if possible).
- The number of any police officer attending the scene.
- The location, time and date of the accident.
- The speed of the vehicles involved.
- The width of the road, details of road markings and signs, the state of the road surface, and the weather conditions.
- Any marks or debris on the road relevant to the accident.
- A rough sketch showing the vehicle positions before and after the accident. It's helpful to make a note of the vehicle positions in terms of distance from fixed landmarks, such as lamp posts, buildings, etc.
- Whether any of the other vehicle occupants were wearing seat belts.
- If the accident occurred at night or in poor visibility, whether vehicle lights or street lights were switched on.
- If you have a camera, take a picture of the vehicles and the scene.

If the other driver refuses to give you his/her name and address, or if you consider that he/she has committed a criminal offence, inform the police immediately.

VAUXHALL ASTRA & BELMONT

ACCIDENT REPORT DETAILS 43

Use this page to record all the relevant details in the event of an accident.
You should transfer all the information recorded on this page onto the Motor Vehicle Accident Report form which you can obtain from your insurance company.

OTHER DRIVER DETAILS

FULL NAME OF DRIVER:

ADDRESS:

POST CODE:

HOME TELEPHONE: WORK TELEPHONE:

DRIVING LICENCE NUMBER: ISSUED:

DATE OF BIRTH: DATE DRIVING TEST PASSED:

TYPE OF LICENCE: (tick box) Full ☐   Provisional ☐   Heavy Goods ☐

PERMITTED GROUPS:

FULL NAME OF OWNER:

ADDRESS:

POST CODE:

TELEPHONE:

INSURANCE COMPANY:

POLICY NUMBER:

AGENT OR BROKER:

TELEPHONE:

OTHER VEHICLE DETAILS

MAKE: MODEL:

YEAR: CC:

REGISTRATION NUMBER:

ACCIDENT REPORT DETAILS

CIRCUMSTANCES OF ACCIDENT

DATE: TIME: am/pm

PLACE: (Street or Road)

TOWN: COUNTY:

SPEED: WERE THE POLICE CALLED? Yes ☐ No ☐

If 'Yes' give details of Police Constabulary concerned:

DETAILS OF WHAT HAPPENED (Please use the page opposite to make a rough sketch map)

WHAT WAS THE WEATHER LIKE?

INDEPENDENT WITNESS 1

TELEPHONE:

INDEPENDENT WITNESS 2

TELEPHONE:

DETAILS OF ANY INJURIES SUSTAINED BY EITHER PARTY:

VAUXHALL ASTRA & BELMONT

ACCIDENT REPORT SKETCH MAP   45

Essential items to include in your sketch map:
- The layout of the road and its approaches.
- The directions and identities of both vehicles.
- Their relative positions at the time of impact.
- The road signs and road markings.
- Names of the Streets and Roads.

ACCIDENTS AND EMERGENCIES

HOW TO COPE WITH A FIRE

Fire is unpredictable, and particularly dangerous when cars are involved because of the presence of petrol, which is highly inflammable.

It's sensible to carry a fire extinguisher on board your car, but bear in mind that the average car fire extinguisher is suitable for use only on the smallest of fires.

In the event of your car catching fire:

- **Switch off the ignition.**
- **Get any passengers and yourself out of the car and well away from danger.**
- **Impose an immediate 'No Smoking' ban.**
- **DO NOT expose yourself, or anyone else, to unnecessary risk in an attempt to control the fire.** Minor fires can be controlled using a suitable extinguisher, but always use an extinguisher at arms-length, and bear in mind that a small vehicle fire can develop into a serious situation without warning.
- **Call for assistance, if necessary** – refer to 'How to cope with an accident' on page 39.

HOW TO COPE WITH A BROKEN WINDSCREEN

There are a number of national specialist companies who offer roadside assistance to drivers who suffer broken windscreens, and it's a good idea to carry the 'phone number of a suitable specialist in your car for use in such situations. Some car insurance policies enable the use of such companies at preferential rates, and it's worth enquiring about this when obtaining insurance quotes.

- **Most of the windscreens fitted to modern cars are of the laminated type.**
Laminated windscreens are made from layers of glass and plastic (usually two layers of glass sandwiching a layer of plastic) which prevents the windscreen from shattering, and preserves the driver's vision in the event of the screen suffering an impact.

   A sharp impact may crack the outer glass layer, but clear vision is usually maintained.

   Obviously, if the windscreen is damaged, it should be replaced at the earliest opportunity, but there is no need to curtail your journey unless the damage is particularly serious.
- **Some older cars may be fitted with toughened windscreens.** If a toughened windscreen suffers an impact, the glass will normally stay intact, but severe 'crazing' will usually occur which is likely to seriously affect the driver's vision.

   If a toughened windscreen breaks, stop immediately, and seek assistance. **DO NOT** attempt to knock the broken glass out of the windscreen frame, as this is likely to result in injury to yourself, and damage to the car. Driving without a windscreen is extremely dangerous and should not be attempted.

WHAT TO DO IF YOUR CAR IS BROKEN INTO OR VANDALISED

If you are unfortunate enough to have your car broken into or vandalised, do not attempt to drive your car away from the scene unless you are satisfied that no damage has been caused which could affect the car's safety.

Where possible notify the police before moving your car, and inform your insurance company at the earliest opportunity.

BREAKDOWNS

In the unfortunate event of a breakdown, the first priority must always be safety – never risk causing an accident by attempting to move the car or trying to work on it if it's in a dangerous position.

Joining one of the national motoring organisations such as the AA or the RAC can provide you with a recovery and assistance service should you break down. It may also save you money, as local garages often charge high rates to provide a recovery service.

This Section provides advice on the course of action to follow should you break down. There are also Sections covering the procedure to follow for towing, and on how to cope with two of the more common causes of breakdowns – a puncture, and a flat battery.

Breakdowns on an ordinary road (not a motorway)

Most importantly of all, common sense must be used, but the following advice should prove helpful.

- **Warn approaching traffic,** to minimise the risk of a collision.
  Where possible, switch on the car's hazard warning flashers, and at night leave the sidelights switched on.
  If a warning triangle is carried, position it a reasonable distance away from the scene of the breakdown, to give approaching drivers sufficient warning to enable them to slow down. Decide from which direction the approaching traffic will have least warning, and position the triangle accordingly. Place the warning triangle on the road surface, out from the edge of the road, where it can be easily seen.
  If possible, send someone to warn approaching traffic of the danger which exists, and to encourage the traffic to slow down.
- **Get the passengers out of the car** as a precaution should the car be hit by another vehicle. The passengers should move well away from the car and approaching traffic, up (or down) an embankment, or into a nearby field for example.
- **If possible, move the car to a safe place.** If the car cannot be safely driven or pushed to a place of safety, call a recovery service to provide assistance. *Don't expose yourself or anyone else to unnecessary danger in order to move the car.*
- **If the car can't be safely moved, turn the steering wheel towards the side of the road.** If the car is hit, it will then be pushed into the side of the road, not into the path of traffic.
- **Try to find the cause of the problem.** Only attempt this if the car is in a safe position away from traffic. Refer to *'Fault finding'* on page 143 for details.
- **If necessary, call for assistance** from one of the national motoring organisations if you're a member, or from a local garage who can provide a recovery service.

VAUXHALL ASTRA & BELMONT

Motorway breakdowns

In the event of a motorway breakdown, due to the speed and amount of traffic there are a few special points to consider, and the following advice should be followed.

- **If possible, switch on the car's hazard warning flashers, and at night leave the sidelights switched on.**
- **DO NOT open the doors nearest to the carriageway, and DO NOT stand at the rear of the car, or between it and the passing traffic.**
- **Get the car off the carriageway and onto the hard shoulder as quickly as possible, and as far to the left as possible** – never forget the danger from passing traffic. Turn the steering wheel to the left, so that the car will be pushed away from the carriageway if hit from behind.
- **If you can't move the car off the carriageway** – get everyone out of the car and well away from the carriageways (up/down an embankment, or into a nearby field for example) as quickly as possible. Never forget the danger from passing traffic.
  Call for assistance as described below.
  Attempt to warn approaching traffic of your car's presence by signalling from well to the back of the hard shoulder, NOT the carriageway. DO NOT put yourself at risk. Face the approaching traffic, and be prepared to move quickly clear of the hard shoulder onto the verge if any vehicles pull onto the hard shoulder.
- **Get the passengers out of the car** – as a precaution should the car be hit by another vehicle. The passengers should move well away from the car and approaching traffic, up (or down) an embankment, or into a nearby field for example. If animals are being carried, it may be advisable to leave them in the vehicle. In any case, animals and children **must** be kept under tight control.
- **Call for assistance** – using the nearest emergency telephone on your side of the carriageway. NEVER cross the carriageways to use the emergency telephones.
  The direction of the nearest telephone is indicated by an arrow on the marker post behind the hard shoulder (the marker posts are positioned 100 metres/300 feet apart).
  Each telephone is coded with details of its location, and the operator will automatically be informed of your position on the motorway when you make the call.
  Give the operator details of your car type and registration number and, if you're a member of one of the motoring organisations, ask the operator to arrange assistance.
- **Don't leave your car unattended for a long period.**

Changing a wheel

- **First of all, make sure that the car is in a safe position.** It may be safer to drive the car slowly forwards and risk damaging the wheel, than to risk causing an accident.
- **If you're in any doubt** as to whether you can safely change the wheel without putting yourself or other people at risk, call for assistance.
- **All the tools required to change a wheel, together with the spare wheel, are located beneath a cover in the luggage compartment.** The spare wheel is secured by a plastic wing nut. However, if you're carrying any luggage, you'll need to move it first for access to the spare wheel. On Estate models, the cover is hinged and must be released by turning the fastener. The tools are located in a bag to the rear of the spare wheel, and on Estate models the

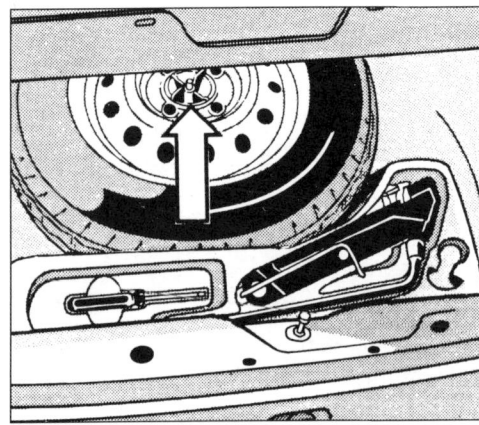

▲ *Spare wheel and tools location on Hatchback and Saloon models – arrow shows plastic wing nut retaining spare wheel*

VAUXHALL ASTRA & BELMONT

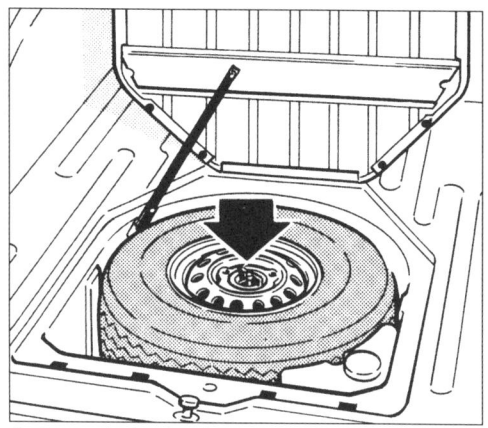

BREAKDOWNS 49

▲ Spare wheel and tools location on Estate models – arrow shows plastic wing nut retaining spare wheel

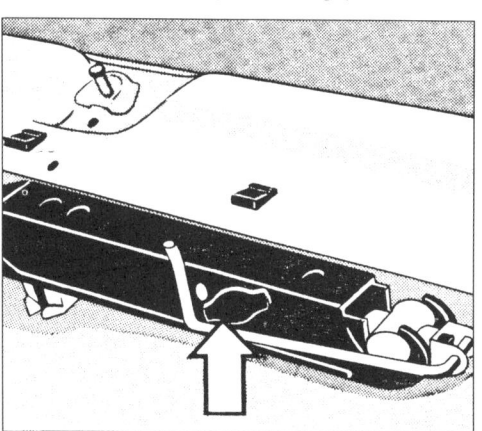

▲ Prising off the hub cap and wheel bolt caps

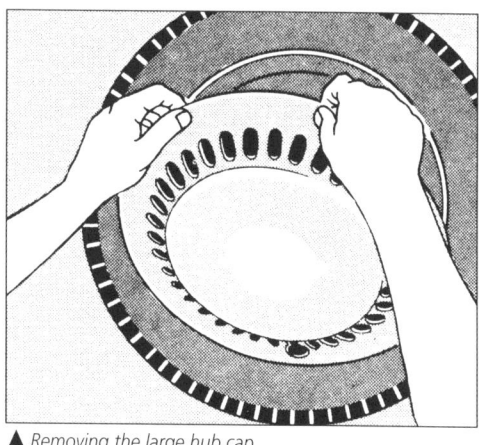

▲ Jack location on Estate models – arrow shows wing-headed bolt

▲ Removing the large hub cap

jack is secured by a wing-headed bolt. Lift the cover and remove the tools, then unscrew the plastic wing nut and lift the spare wheel out of the storage well. Place them on the ground (don't put any luggage back yet because the wheel with the punctured tyre will have to be put back in the spare wheel well).

- **Make sure that the handbrake is applied,** and select reverse gear on models with a manual gearbox, or position **P** on models with automatic transmission.
- **Prise off the wheel bolt caps and/or hub cap** with the screwdriver supplied in the tool kit – on models where the hub cap covers

▲ Loosening the wheel bolts

VAUXHALL ASTRA & BELMONT

BREAKDOWNS

the complete main part of the wheel, use the fingers also to apply extra force to the edge of the cap.
- **Slacken the four wheel bolts** by at least half a turn using the wheelbrace provided. If possible, chock the diagonally-opposite wheel to the one being removed with large stones or pieces of wood.
- **The jacking points** are shown on the accompanying illustrations, and they are indicated on the car by two depressions on the side sills, one in front of the rear wheel and the other behind the front wheel.
- **Choose the jacking point nearest the wheel to be changed,** and position the jack arm directly below it. Turn the jack handle clockwise so that its head moves upwards, and guide it onto the jacking point so that it enters the recess. Make doubly sure that the base of the jack is positioned directly below the jacking point, then continue to turn the handle clockwise.

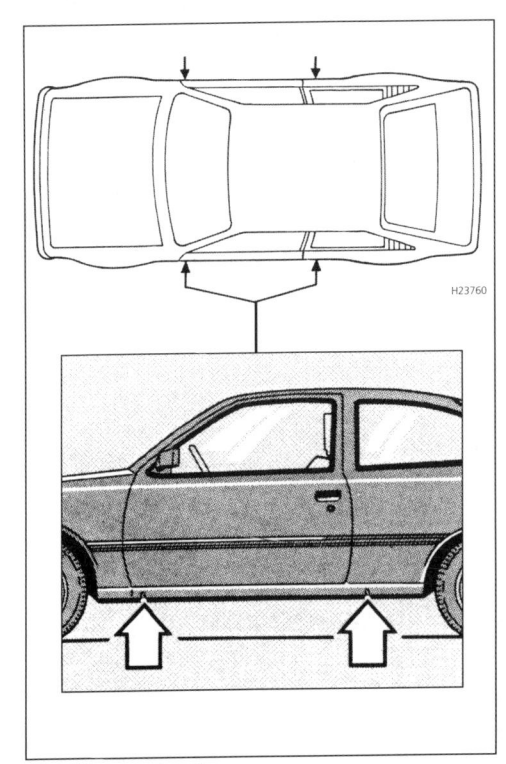

▲ *Using the car jack*

▲ *Jacking point locations*

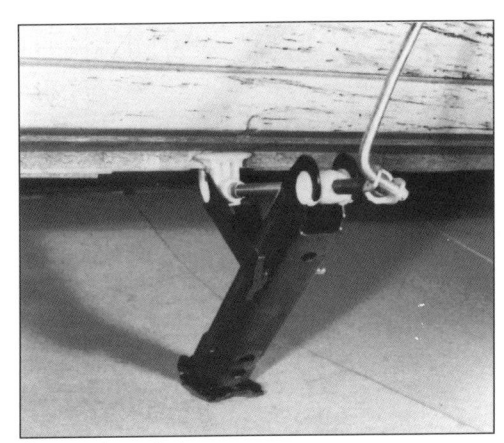

- **If the jack tilts at all or slips,** lower it and start again. Carry on turning the jack handle slowly to raise the car until the wheel is clear of the ground, making sure that the car doesn't move on the jack. Slide the spare wheel under the door sill panel for the time being, as near to the punctured wheel as possible. This will break the car's fall (and minimise the risk of injury to you) if the car slips off the jack or the jack collapses.
- **Unscrew all the wheel bolts,** then remove the punctured wheel. Sometimes it may be necessary to tap the wheel in order to release it, but if this is done, make sure that the car doesn't move on the jack. Remove the spare wheel from under the side of the car, and slide the punctured wheel under there in its place for the time being.
- **Fit the spare wheel to the car,** and lightly tighten the wheel bolts in a diagonal sequence. Slide out the punctured wheel and slowly lower the car to the ground.
- **Remove the jack, and fully tighten the wheel bolts.** There is no need to strain yourself doing this, just make sure that the bolts are tight, using only the wheelbrace provided.
- **Refit the hub cap and/or wheel bolt caps,** giving a sharp tap with the palm of the hand to secure. Where applicable, make sure that the special access hole for the tyre valve is located over the valve.
- **Stow the removed wheel and the tools** in

BREAKDOWNS 51

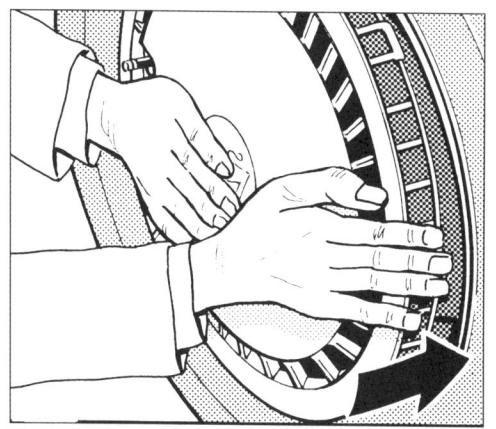

▲ Tap the hub cap sharply to fit it on the wheel

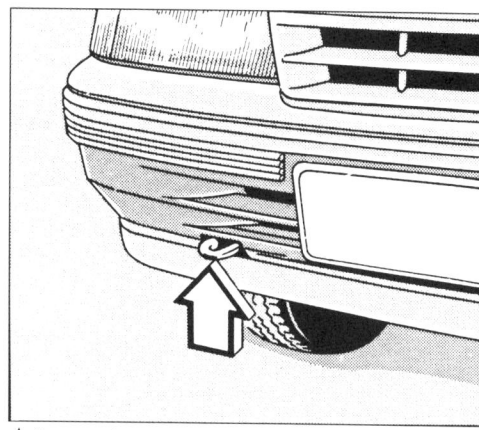

▲ Front towing eye

the spare wheel well or carrier as applicable.
- **Remove the wheel chocks,** and make a final check to ensure that all tools and debris have been cleared from the roadside.
- **Select neutral,** where applicable, before starting up and driving away.
- **Have the puncture repaired as soon as possible**, or another puncture will leave you stranded.

Towing

If your car needs to be towed, or you need to tow another car, special towing eyes are provided at the front and rear of the car.

On some 'Mark 2' models (notably the 'Mark 2' Astra GTE) the front towing eye is detachable, and is located in the spare wheel well – open the flap on the front spoiler, insert the eye in the metal tube, and lock by fitting the special pin.

When being towed, the ignition key should be turned to position **II** (steering lock released, and ignition warning light on). This is necessary for the direction indicators, horn, brake lights and wipers to work.

Note that when being towed, the brake servo does not work (because the engine is not running). This means that the brake pedal will have to be pressed harder than usual to operate the brakes, so allowance should be made for greater stopping distances. On models fitted with power steering, greater effort will be necessary to steer the car, since

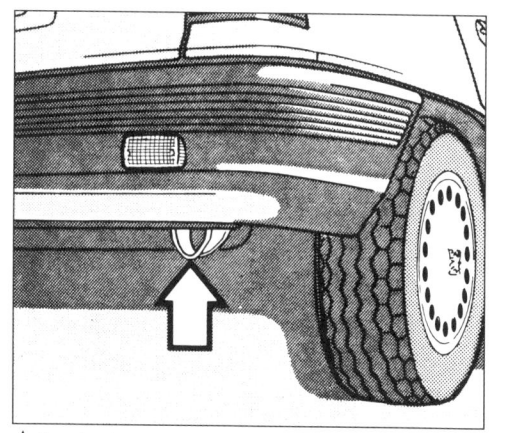

▲ Rear towing eye

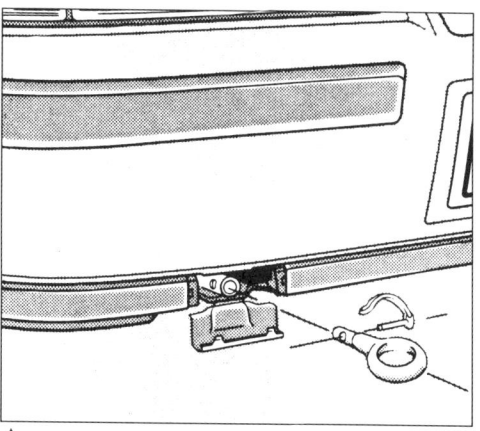

▲ Front towing eye on some 'Mark 2' models

VAUXHALL ASTRA & BELMONT

BREAKDOWNS

the power steering pump will not be operating if the engine is stopped. If the breakdown doesn't prevent the engine from running, the engine could be started and allowed to idle so that the servo (and power steering where fitted) functions while towing.

On manual gearbox models, the gear lever should be positioned in neutral.

When a car with an automatic transmission is being towed, the gear selector lever should be moved to the 'N' position. To avoid damage to the transmission, do not tow the car any faster than 50 mph (80 km/h) or any further than 60 miles (100 km). Ideally, a car with automatic transmission should only be towed with the front wheels off the ground (where this can be arranged), especially if a transmission problem is suspected.

An 'On tow' notice should be displayed prominently at the rear of the car on tow to warn other drivers.

Make sure that both drivers know details of the route to be taken before moving off, as it is difficult, and dangerous, to try to communicate once under way.

Before moving away, the tow car should be driven slowly forwards to take up any slack in the tow rope.

The driver of the car on tow should make every effort to keep the tow rope tight at all times, by gently using the brakes if necessary, particularly when driving downhill.

Drive smoothly at all times, particularly when moving away from a standstill. Allow plenty of time to slow down and stop when approaching junctions and traffic queues.

Starting a car with a flat battery

Apart from old age (most batteries should last for at least three years), a battery will normally only go flat if there is a fault in the charging circuit (indicated by a continuously glowing ignition warning light), or when a particular circuit (eg headlights) is left on for a long time with the engine switched off.

To start a car with a flat battery, you can either use a push or tow start (models with manual gearbox only), or a set of jump leads.

Note, however, that where a maintenance-free battery is fitted, certain precautions must

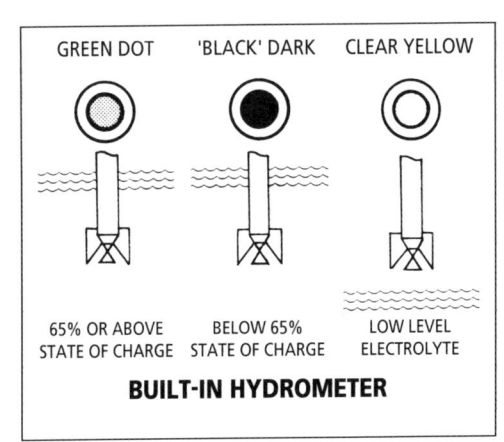

▲ *Battery condition indicator on maintenance-free battery*

be observed. Batteries of this type were fitted extensively to the Astra/Belmont range as original equipment, and will typically be marked 'Freedom Battery'. If the battery is of this type, look carefully on the top of the battery for a small round window – this indicates battery condition by changing colour, as shown in the accompanying illustration.

Do not attempt to start the engine if the battery condition indicator is clear or yellow in colour. This means that the electrolyte level in the battery is too low to allow further use, and the battery should be replaced. If the indicator is either green (battery in good state of charge) or black (battery needs charging), the engine may be started using any of the following methods.

TOW/PUSH STARTING (MANUAL GEARBOX MODELS ONLY)

Warning: *Due to the possibility of damaging the catalyst unit, models fitted with a catalytic converter should not be tow- or push-started – use the jump lead method described later in this Section.*

The method used to start the engine is the same whether the car is being towed or being pushed but, if the car is to be towed, first read the previous Section on towing for details of where to connect the tow rope. Proceed as follows:

1 Turn the ignition key to position **II** to switch on the ignition and unlock the steering. Switch off all other electrical loads.

2 If the engine is cold, on models with a

BREAKDOWNS 53

manual choke, pull out the choke control. On 1.3, 1.4 & 1.6 litre models with an automatic choke, fully depress the accelerator pedal once then release it; if the engine is warm, hold the accelerator pedal halfway down. On 1.8 & 2.0 litre models, do not depress the accelerator pedal at all.

3 Depress the clutch pedal and select second or third gear. Hold the clutch pedal down.

4 Tow or push the car, then release the clutch pedal. The engine will turn over, and should start (don't worry about the car 'juddering' as the engine turns). If the car is being towed, as soon as the engine starts, depress the clutch pedal and *gently* brake the car so that you don't run into the tow vehicle.

STARTING USING JUMP LEADS

Note: *Starting using jump leads can be dangerous if done incorrectly. If you're uncertain about the following procedure, it's recommended that you ask someone suitably qualified to do the jump starting for you.*

Before attempting to use jump leads, there are a few important points to note.

First of all, use only proper jump leads which have been specifically designed for the job.

Make sure that the booster battery (the fully-charged battery) is a 12-volt type.

Position the two vehicles close enough together to connect the leads, but **do not** allow the vehicles to touch.

Turn off all electrical circuits.

DO NOT allow the ends of the two jump leads to touch at any time during the following procedure. Proceed as follows.

1 Open the vehicle bonnets, and connect one end of one of the jump leads (usually the red one) to the positive (+) terminal of the flat battery, and the other end to the positive terminal of the booster battery.

2 Connect one end of the other jump lead to the negative (-) terminal of the booster battery, and connect the other end to a suitable earth point on the car with the flat battery at least 50 cm (18 in) away from the battery (eg a clean, bare metal area on the engine block or body). Make sure that all the lead clips are secure.

3 Start the engine of the vehicle with the booster battery, and let it run for a few minutes, then start the engine of the car with the flat battery in the normal way.

4 When the engine is running smoothly, disconnect the jump leads in exactly the reverse order to that in which they were connected.

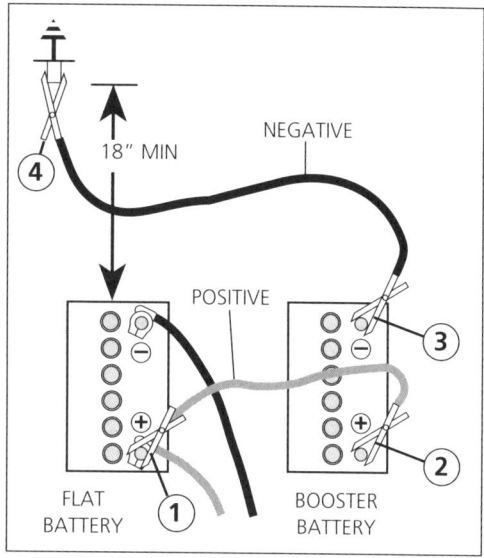

▲ Jump start lead connections for negative earth vehicles – connect leads in order shown

What to carry in case of a breakdown

Carrying the following items may help to reduce the inconvenience and annoyance caused if you're unlucky enough to break down during a journey:

- Alternator drivebelt (refer to *'Servicing'* on page 116 for details of renewal)
- Spare fuses (refer to *'Bulb, fuse and relay renewal'* on page 139 for details of renewal)
- Set of principal light bulbs (refer to *'Bulb, fuse and relay renewal'* on page 131 for details of renewal)
- Battery jump leads
- Tow-rope (refer to *'Towing'* on page 48 for details of towing procedure)
- Litre of engine oil (refer to *'Regular checks'* on page 80 for details of checking oil level)
- Torch

VAUXHALL ASTRA & BELMONT

54 **DRIVING SAFETY**

VAUXHALL ASTRA & BELMONT

DRIVING SAFETY

In the UK, more people are injured or killed every year due to road accidents than through any other single cause.

Car manufacturers are paying more attention to passive safety when designing new cars, but drivers must ultimately take active responsibility of ensuring the safety of themselves, their passengers, and just as importantly, other road users.

There will always be accidents on the roads, but taking the time to read the following advice may help you to avoid such a mishap.

The following Sections aim to give advice which will help you to reduce the risk of becoming involved in an incident when driving, not necessarily by changing the way that you drive, but simply by making you aware of some of the potential risks which can easily be avoided. For more information and practical advice on road safety, it is well worth considering taking part in one of the various courses provided by organisations such as RoSPA and the Institute of Advanced Motorists.

BEFORE STARTING A JOURNEY

As a driver, before beginning a journey it's important to make sure that you're comfortable so that you can concentrate on driving without any unnecessary distractions. Spending a few moments carrying out a few simple checks and adjustments will help to make your journey more relaxing, safer and hopefully trouble-free.

Driver comfort

Before driving, make sure that you're comfortable, and that you can operate all the controls easily, particularly if someone else has recently driven the car.

- **Seat**
  Make sure that the seat is positioned a comfortable distance from the steering wheel and the pedals, so that you can operate all the controls comfortably
- **Seat back**
  Make sure that the seat back is adjusted to give your back plenty of support. Your back should rest against the seat, and there should be no need to lean forwards from the seat back. You should be able to reach the steering wheel comfortably, and you should be able to turn the wheel easily without stretching.
- **Head restraint**
  If a head restraint is fitted, it should be adjusted to support your head in the event of an accident. Your head should not rest against the restraint during normal driving and, as a rough guide, the restraint is correctly positioned when its top is in line with your eyes. Similarly, make sure that any passenger head restraints are adjusted correctly if passengers are being carried.
- **Mirrors**
  All the mirrors should be adjusted so that they give a clear view behind the car without the need to move your head unnecessarily.
- **Steering column**
  If the car is fitted with an adjustable steering column, adjust the column so that the steering wheel can be reached comfortably, and the wheel can be turned easily without stretching.

Checks

If you're planning a long journey, refer to *'Regular checks'* on page 77, and make sure that you carry out all the checks mentioned before setting off. Also make sure that you know where the spare wheel and the tools required for wheel changing are located (refer to *'Breakdowns'* on page 48).

DRIVING SAFETY

DRIVING IN BAD WEATHER

When driving in bad weather, always be prepared. Bad weather should never take you by surprise. Listen to a weather forecast: you should always be aware of the possibility of poor conditions on your intended journey.

Driving in bad weather requires more concentration, and is more tiring than driving in good weather conditions. Never allow yourself to be distracted by talkative passengers or loud music in the car and, if you feel tired, stop at the next opportunity and take a break.

Always look well ahead, so that you're aware of the condition of the road surface, and any obstacles; **slow down if necessary**.

The following advice is intended to help you to drive more safely in various bad weather conditions. However, above all, use common sense.

Rain

- Use dipped headlights in poor visibility.
- Slow down if visibility is poor or if there is a lot of water on the road surface.
- Keep a safe distance from the vehicle in front (stopping distances are doubled on a wet road surface).
- Be particularly careful after a long period of dry weather. Under these conditions, rain can make the road surface very slippery.
- Don't use rear foglights unless visibility is **seriously** reduced (generally less than 100 metres/300 feet). Foglights can dazzle drivers following behind, especially in motorway spray.

Fog

- Slow down. Fog is deceptive, and you may be driving faster than you think. Fog can also be patchy, and the visibility may suddenly be reduced. Always drive at a speed which allows you to stop in the distance you can see ahead.
- Use dipped headlights. Using main-beam headlights will usually reduce the visibility even further, as the fog will scatter the light.
- Use foglights (where they are fitted) if it's genuinely foggy (generally, where visibility is reduced to less than 100 metres/300 feet), and not just misty.
- Keep a safe distance from the vehicle in front.
- Use your windscreen wipers to clear moisture from the windscreen.

Snow

- Don't start a journey if there is any possibility that conditions may prevent you from reaching your destination. Listen to a weather forecast before setting off.
- Before starting a journey, clear **all** snow from the windscreen, windows and mirrors. Don't just clear a small area big enough to see through.
- Slow down.
- Keep a safe distance from the vehicle in front (stopping distances can be trebled or even quadrupled on a snow-covered or icy surface).
- Drive smoothly and gently. Accelerate gently, steer gently and brake gently.
- Don't brake and steer at the same time – this may cause a skid (refer to 'Skid control' on page 60).
- When moving away from a standstill, or manoeuvring at junctions, etc, in a car with a manual gearbox use the highest possible gear that your car will accept to move away, and change up to a higher gear earlier than

VAUXHALL ASTRA & BELMONT

DRIVING SAFETY

usual. In a car with automatic transmission move the selector lever to prevent the transmission from using all the gears (usually position '2' – refer to *'Controls and equipment'* on page 23). This will provide more grip and will help to reduce wheelspin.
- Use main roads and motorways where possible. Major roads are likely to have been 'gritted' and are usually cleared before minor roads.
- Use dipped headlights in poor visibility.
- Don't use rear foglights unless visibility is **seriously** reduced (generally less than 100 metres/300 feet) – foglights can dazzle drivers following behind.

Ice and frost
- Follow the advice given for driving in snow.
- Be prepared for 'black ice'. Although you can't see 'black ice', reduced road noise from the tyres will usually tell you it's there.

Severe winter weather

The best advice in severe winter weather is to stay at home but, if you must drive your car, in addition to following the advice given for driving on snow and ice, there are a few additional pieces of advice which you should bear in mind before setting out on a journey.

- Make sure that you tell someone where you're going, and tell them roughly what time you're expecting to arrive at your destination, and what route you're taking.
- Keep a can of de-icer fluid, a scraper, a set of jump leads and a tow rope in the car at all times.
- Carry plenty of warm clothes and blankets.
- Make sure that you have a full tank of petrol before starting your journey. This will allow you to keep the engine running, without fear of running out of petrol, to provide warmth through the car's heating system should you be delayed by conditions, or worse still stuck.
- Pack some pieces of old sacking or similar material, which you can place under the driving wheels to give better traction if you get stuck.
- Carry a shovel, in case you need to dig yourself or someone else out of trouble.

MOTORWAY DRIVING

Motorway driving requires a great deal of concentration and awareness. Motorways are generally busier than ordinary roads, and the traffic moves faster, so there is less time to react to any changes in road conditions ahead.

This Section covers the fundamental rules for safe motorway driving, and will help you to avoid many of the problems which occur on today's busy motorways. Remember that lack of common sense is probably the biggest cause of accidents on motorways.

Full rules and regulations for motorway driving can be found in 'The Highway Code'.

Joining and leaving a motorway

When you join a motorway, you will normally approach from a 'slip-road' on the left. You must give way to traffic already on the motorway. Watch for a safe gap in the traffic, and adjust your speed in the acceleration lane so that when you join the left-hand lane of the motorway you're already travelling at the same speed as the other traffic. Indicate before pulling onto the motorway. After joining the motorway, allow yourself some time to get used to the speed of the traffic before overtaking.

When you leave a motorway, indicate in plenty of time, and reduce your speed before you enter the slip-road – some slip-roads and

DRIVING SAFETY

link-roads between motorways have sharp bends which can only be taken safely by slowing down. Be very cautious immediately after leaving a motorway, as it can be difficult to judge speed after a long period of fast driving.

Rules for safe motorway driving

- **Concentrate and think ahead** – Always be prepared for the unexpected, and look well ahead. You should be aware of all the traffic in front and behind, not just the vehicle in front of you. If you're concentrating properly, nothing should take you by surprise on a motorway.
- **Always drive at a speed to suit the road conditions** – Don't break the speed limit, especially temporary speed limits which may apply to contra-flow systems or roadworks.
- **Slow down** in bad weather conditions. Driving too fast in fog and motorway spray is a major cause of motorway accidents.
- **Always keep a safe distance from the vehicle in front** – The closer you drive to the vehicle in front, the less chance you have of avoiding an accident if something happens ahead. Remember that you need more space in bad weather conditions.
- **Think** – If the vehicle ahead suddenly stops, do you have enough space to stop without hitting it?
- **Use your mirrors regularly** – It's just as important to be aware of what's happening behind you as it is in front.
- **Always signal your intentions clearly and in plenty of time** – Check your mirrors first, signal *before* you move, and change lanes smoothly and in plenty of time. Don't make any sudden moves which are likely to affect traffic approaching from behind.
- **Keep to the left** – The left-hand lane is *not* the slow lane, and you should drive in this lane whenever possible. You can stay in the middle lane when there are slower vehicles in the left-hand lane, but return to the left-hand lane when you've passed them. The right-hand lane is for overtaking only: it is *not* for driving at a constant high speed. If you use the right-hand lane, move back into the middle lane and then into the left-hand lane as soon as possible, but without cutting in.
- **Take note of direction signs** – You may need to change lanes to follow a certain route, in which case you should do so in plenty of time.
- **Stop as soon as practicable if you feel tired** – Driving on a motorway when feeling tired can be extremely dangerous. If necessary, wind down the window for fresh air. If you feel tired or drowsy, stop at the next service station, or turn off the motorway at the next junction and take a break before continuing your journey. You must not stop on the hard shoulder of a motorway other than in an emergency.
- **If you break down** – Refer to *'Breakdowns'* on page 48.

TOWING A TRAILER OR CARAVAN

When towing a trailer or a caravan, there are a few special points to bear in mind. There's more to towing a trailer or caravan than just simply bolting on a towbar and hitching up!

The law

Before using your car for towing, make sure that you're familiar with any special legislation which may apply, particularly if you're travelling abroad.

Make sure that you know the speed limits applicable for towing.

In some countries, including the United Kingdom, there is a legal requirement to have

DRIVING SAFETY

a separate warning light fitted in the car to show that the trailer/caravan direction indicator lights are working.

Always check on current regulations before you travel.

Before starting a journey

Before attempting to use your car for towing, there are one or two points which should be considered.

- **Make sure that your car can cope with the load which you're towing**
- **Don't exceed the maximum trailer or towbar weights for the car** – Refer to *'Dimensions and weights'* on page 13.
- **Engine** – Don't put unnecessary strain on your car's engine by trying to tow a very heavy load. Obviously, the smaller the engine, the less load the car can comfortably tow. Also bear in mind that the extra load on the engine when towing means that the engine's cooling system may no longer be adequate. Some manufacturers provide modified cooling system components for towing, such as larger radiators, etc, and if you intend to use your car for towing regularly, it may be worth enquiring about the availability and fitting of these components.
- **Suspension** – Towing puts extra strain on the suspension components, and also affects the handling of the car, since standard suspension components aren't usually designed to cope with towing. Heavy duty rear suspension components are available for most cars, and it's worth considering having these fitted if you intend to use your car for towing regularly. Special stabilisers are also available for fitting to most cars to reduce pitching and 'snaking' movements when towing.
- **Make sure that you can see behind the trailer/caravan using the car's mirrors** – Additional side mirrors with extended arms are available to fit most cars. The mirrors must be fitted with folding arms (so that they fold if hit), and should be adjusted to give a good view to the rear at all times. The mirrors can be removed when the car isn't being used for towing.
- **Make sure that the tyre pressures are correct** – Unless a light, unladen trailer is being towed, the car tyres should be inflated to their 'full load' pressures. Also make sure that where applicable the trailer or caravan tyres are inflated to their recommended pressure.
- **Make sure that the headlights are set correctly** – Check the headlight aim with the trailer/caravan attached, and if necessary adjust the settings to avoid dazzling other drivers.
- **Make sure that the trailer/caravan lights work correctly** – Make sure that the tail lights, brake lights, direction indicator lights and, where applicable, the reversing lights and fog lights all work correctly. Note that the additional load on the direction indicator circuit may cause the lights to flash at a rate slower than the legal limit, and in this case you may need to fit a 'heavy duty' flasher unit.
- **Make sure that the trailer/caravan is correctly loaded** – Refer to the manufacturer's recommendations for details of loading. As a general rule, distribute the weight so that the heaviest items are as near as possible to the trailer/caravan axle. Secure all heavy items so that they can't move. To provide the best possible control and handling of the car when towing, the manufacturers recommend an optimum noseweight for the trailer/caravan when loaded (refer to *'Dimensions and weights'* on page 13 for details). The noseweight can be measured using a set of bathroom scales as follows:

  Place a stout piece of wood between the trailer/caravan tow hitch cup and the scales platform, and read off the weight with the trailer/caravan level. If necessary, redistribute the load, to arrive as close as possible to the recommended noseweight. **Do not** exceed the maximum recommended noseweight.

Driving tips

Towing a trailer or caravan will obviously affect the handling of the car, and the following advice should be followed when towing.

- **If possible, avoid driving with an unladen car and a loaded trailer/caravan** – The uneven weight distribution will tend to make the car unstable. If this is unavoidable, drive slowly to allow for the instability.

DRIVING SAFETY

- **Always drive at a safe speed** – The stability of the car and the trailer or caravan decreases as the speed increases. Always drive at a speed which suits the road and weather conditions, and always reduce speed in bad weather and high winds – especially when driving downhill. If the trailer or caravan shows any sign of 'snaking', reduce speed immediately – never try to stop 'snaking' by accelerating.
- **Always brake in good time** – If you're towing a trailer or caravan which has brakes, apply the brakes gently at first, then brake firmly. This will help to prevent the trailer wheels from locking. In cars with a manual gearbox, change into a lower gear before going down a steep hill so that the engine can act as a brake, and similarly, in cars with automatic transmission, move the selector lever to position '2', or '1' in the case of very steep hills.
- **Don't change to a lower gear unnecessarily** – Unless the engine is labouring, stay in as high a gear as possible to keep the engine revs as low as possible. This will help to avoid the engine overheating.

ALCOHOL AND DRIVING

By far the best advice on drinking and driving is **DON'T**.

You may feel fine, but it's a proven fact that even one small alcoholic drink will impair your driving to some extent.

Drinking alcohol has the following effects:

- **Reduces co-ordination**
- **Increases reaction time**
- **Impairs judgement of speed, distance and risk**
- **Encourages a false sense of confidence**

The risk of an accident increases sharply after drinking alcohol, and approximately one third of the total number of people killed in road accidents each year have blood alcohol levels above the legal limit for driving. Remember that you're putting other people as well as yourself at risk if you drive after drinking.

The legal limit for blood alcohol level in the UK is 80 milligrams per 100 millilitres. This doesn't correspond to any particular quantity of drink, as the amount of drink required to reach this level varies from person to person. The driving of many people who feel perfectly sober is seriously affected well below the legal limit.

The penalties for driving over the legal limit are severe, and can mean losing your driving licence, a heavy fine, or imprisonment. It's also important to realise that the laws abroad can be far more severe than in the UK, and some countries have a total ban on driving after drinking alcohol.

- **The safest course of action is not to drink and drive.**

SKID CONTROL

If you drive sensibly with due regard for the road conditions, you should never find yourself in a situation where your car is skidding. The following advice will help you to avoid situations which may cause a car to skid, and explains how to regain control of your car quickly should the need arise.

VAUXHALL ASTRA & BELMONT

DRIVING SAFETY

A skid is caused by one or a combination of the following:

- **Excessive speed in relation to the road conditions.**
- **Harsh or excessive acceleration.**
- **Sudden or excessive braking.**
- **Coarse or excessive steering.**

The most common basic cause of skidding is rough handling of the car's controls. Therefore the key to safe driving and preventing a skid is smooth, gentle handling of the car and its controls. Try to apply smooth pressure to the brake pedal, accelerator and steering wheel rather than just suddenly moving them a certain distance.

When a car is skidding, the following things happen:

- **The car is out of control. You can take action to regain control, but while skidding, the car can't be fully controlled.**
- **A car with the front wheels skidding can't be steered.**
- **A car with the rear wheels skidding is likely to spin round if any steering lock is applied.**
- **A car with all four wheels skidding will continue in a straight line in the direction it was travelling when the skid started, regardless of which way the car is pointing.**

Prevention is better than cure. When approaching a corner, reduce speed early by smooth progressive braking **in a straight line**. In a car with a manual gearbox, brake to an appropriate speed and select the correct gear for the corner. Turn into the corner early and smoothly, gently increasing the steering lock if necessary, and corner with your foot gently on the accelerator, so that the engine just 'pulls' the car through the corner, maintaining a constant speed. As you leave the corner, accelerate gently and smoothly.

If you follow the above advice, you should avoid finding yourself in a situation where your car is skidding. However, should you be unfortunate enough to experience skidding, the basic rule for skid control is to **remove the cause** of the skid. If you cause a skid by accelerating, braking or steering, **ease off**, and, when the skid stops, re-apply the accelerator, brake or steering control, but this time more gently and smoothly. If the front wheels are skidding, to regain steering control quickly, steer 'into the skid'. This is often misunderstood, and it basically means look in the direction you want to be facing, and steer in that direction – and be prepared to reduce the steering quickly if necessary when the wheels regain their grip.

Several organisations offer courses in skid control, using specially-equipped cars under carefully-controlled conditions. One of these courses could prove to be a very worthwhile investment, possibly helping you to avoid an accident.

ADVICE TO WOMEN DRIVERS

Unfortunately, it's a fact that women are more likely than men to be the subject of unwelcome attention when driving alone.

There's no reason to think that driving alone spells trouble, but it's a good idea to be aware of certain precautions which will help to avoid finding yourself in an unpleasant situation.

This Section aims to give advice which will help to reduce any risks when driving alone, and will help you to deal with the situation, should you be unfortunate enough to find yourself the subject of unwelcome attention.

Several organisations (some local police authorities, AA, etc) run courses especially for women drivers, in which subjects such as basic car maintenance and self defence, etc, are covered.

Sensible precautions
- **The car**

Make sure that you're familiar with your car.

Make sure that your car is regularly serviced. Refer to *'Regular checks'* on page 77, and make sure that all the checks described are carried out regularly. This will minimise the possibility of a breakdown.

Make sure that you know how to change a wheel, and make sure that the car jack is in good condition in case you have a puncture - refer to *'Breakdowns'* on page 48.

DRIVING SAFETY

● The journey

If you're planning to undertake a particularly long or unfamiliar journey, the following advice may be helpful.

If you're travelling on an unfamiliar route, make a few notes before you set off, reminding yourself of the road numbers, where to turn, which junctions to use on motorways, etc. Try to stick to main roads where possible. Always carry a map.

Make sure that you have enough petrol for the journey. If you need to fill up during the journey, make sure that you do so in plenty of time, before the fuel level gets too low, and before petrol stations close if travelling at night.

If you think that your family or friends may be worried, you may want to 'phone someone at your destination before setting off to tell them that you're leaving, and tell them what time you expect to arrive. Similarly, when you arrive, you may want to 'phone someone at your starting point to confirm that you've arrived safely.

● Driving in traffic

Avoid attracting unnecessary attention.

Avoid 'jokey' stickers in car windows, which may encourage unwelcome attention.

Avoid eye contact with 'undesirables'.

If you're being followed, pull over and slow down, but **don't** stop. **Don't** drive to your home, but find a busy and well lit place. If the person following persists, blow your horn and flash your lights to attract attention.

If you're stopped by traffic or another vehicle, lock the doors and close the windows. **Don't** ram the other vehicle - damage to your car might prevent your escape.

● Leaving your car

Don't leave any clues that the car is being driven by a woman; make sure that any 'feminine' items are hidden from sight before leaving your car.

Refer to *'Car crime prevention'* on page 71 and take note of the advice given.

Try to park in a well-lit, preferably busy area.

If you park in a car park, try to park close to an exit, or close to the attendant's station.

Always reverse into a parking space, so that you can drive away quickly if necessary after returning to your car.

Take note of any 'landmarks' so that you can find your car quickly when you return.

Always lock your car.

When you return to your car, walk with a group of people, if possible. Have the keys ready so that you don't have to spend unnecessary time outside your car searching for them, which may attract attention.

Before getting into your car, briefly check for forced entry, and look into the car for any suspicious signs. **Don't** get into the car if you notice anything suspicious.

● Breakdowns

Remember that you're more likely to be injured in an accident than a personal attack.

Refer to *'Breakdowns'* on page 47 and take note of the advice given but, in addition, bear in mind the following advice.

Always walk to 'phone for assistance, **don't** accept a lift. When you 'phone for assistance, mention that you're an unaccompanied woman.

If someone stops while you're 'phoning for help, give the operator details of the other vehicle's registration number and a brief description of the car and driver. If the driver approaches you, tell them that you have passed on his/her details to the police. If the driver's intentions are honourable, your reaction will be understood.

If you decide to stay with, but outside the car, leave the nearside door unlocked and slightly open, so that you can get inside quickly to lock yourself in if necessary.

If you decide to stay inside the car, sit in the passenger seat and lock the door - this will give the appearance that you're accompanied.

When help arrives, ask for some form of identification (even from a policeman) before giving any of your own details.

If someone offers assistance, tell them the police have been informed and are arranging recovery. If you have not yet contacted the police, consider asking the person to do so on your behalf, **but** if you're uncertain about the person's intentions, tell them that the police are aware and ask them to call the police again for you.

VAUXHALL ASTRA & BELMONT

DRIVING SAFETY

● Accidents

Keep calm.

Refer to *'Accidents and emergencies'* on page 39 and take note of the advice given but, in addition, bear in mind the following advice.

If you're bullied or shouted at, lock yourself in your car (if it's safe to do so), and wind the windows up. Communicate through a small gap at the top of the window.

If things get out of hand, refuse to talk to anyone except the police.

Self defence

This is a last resort and should not be used unless all else has failed.

There is no substitute for taking a properly supervised course in self-defence, but we aim to outline a few of the basic moves below.

If you are approached, act in a composed manner and use a soft tone.

In the event of an attack, stay calm - you can still safeguard yourself.

- **If you're held by one arm** - Use your free hand to grasp the attacker's thumb and twist the thumb sharply back towards the wrist.
- **If you're facing your attacker** - If your shoulders are held, thrust your hands/arms upwards and diagonally outwards to dislodge the attacker's grip.

  Use your knees or the blade of your hand to chop your attacker's groin, your feet to kick the shins, and your fingers to poke the eyes.
- **If you're attacked from behind** - Quickly lean forward, twisting your head sideways into the attacker to keep your airway (for breathing) clear. Often this will cause the attacker to lose balance.
- **If you've been pulled backwards already** - Turn your head as above, and chop hard with the blade of your hand or a clenched fist to the groin, or an elbow in the stomach.
- **If you've been forced to the floor** - Use your feet and legs to kick against the attacker's shins whilst swivelling your body to keep the attacker at bay.

Be ready to run as soon as the attacker's grip is released, and SCREAM!

CHILD SAFETY

Every day, thousands of parents strap their offspring into child car seats, confident that they have done everything possible to protect their loved ones. However, a recent survey has shown that many young children are travelling in car seats which have been incorrectly fitted or are being incorrectly used, making them potentially dangerous should they be involved in an accident. The following advice will help you to make sure that you take every possible precaution to ensure the safety of your children when driving.

How to avoid unnecessary risks

- **Never allow young children to travel in a car unrestrained,** even for the shortest of journeys.
- **Never carry a child on an adult's lap or in an adult's arms.** Although you may feel that your baby is safer in your arms, this is not the case. An adult holding a child is far more likely to cause injury to the child than to give protection in the event of an accident.

VAUXHALL ASTRA & BELMONT

DRIVING SAFETY

- **Never rely solely on an adult seat belt to restrain a child,** and never sit a child on a cushion to enable a seat belt to fit properly.
- **Always strap young children into a properly designed child car seat** when carrying them in a car. The cost of a good car seat is a very small price to pay to save your child's life.

Choosing a child car seat

- It's vitally important to choose the correct type of car seat for your child. A wide range of child car seats is available, and it's worth spending some time looking at the various seats on the market before deciding on the most suitable seat for your particular requirements.
- Although age ranges are often given by the manufacturers, these should be taken as a rough guide. It's the weight of the child which is important; for example, a smaller than average baby could use a baby car seat for longer than a heavy baby of the same age.
- **Never** buy a secondhand child seat. This may sound like a ploy from the manufacturers to sell more seats, but all too often a secondhand seat is sold without the instructions, and sometimes there are parts missing. This often leads to secondhand seats being incorrectly fitted. A secondhand seat may have been damaged or weakened through carelessness or misuse, without necessarily showing any visible signs until it's put to the test in an accident!
- Your baby's first contact with the outside world is often on that first ride home from hospital, so make sure that you're prepared, and buy a car seat before your baby is born.
- Some seats are designed for babies from birth to a weight of around 22 pounds (10 kg), which for most babies is around nine months old. Usually, these first car seats are light and easy to transport through the use of a handle. This means that a sleeping baby can be carried from the car into the house without waking. Some baby seats come complete with a built-in headrest, which is very important for the early weeks when a baby's head needs to be supported.

Certain seats can be fitted so that the baby faces the back of the car (ie rearward facing seats). This may be considered to improve safety, as the baby is supported across the back rather than purely by the harness, if the car is involved in a frontal impact. Combination type seats can then be used forward facing when the child reaches approximately nine months old.

Other types of seat may be designed to accommodate children up to around 40 pounds (18 kg) or approximately four years old.

Alternatively, some seats use the car's seat belts to hold both the child seat and the child, and these seats are obviously easier to fit. Make sure that this type of seat is fitted with a seat-belt lock, so that the seat belt cannot be pulled out of place or slackened. Another advantage of these seats is that they can easily be transferred from car to car.

- Try to choose a seat which has an easily adjustable harness, as this makes it easier to ensure that the harness fits the child securely for each trip. If the harness can be quickly loosened, it makes getting a struggling child in and out a little easier.

Using a child car seat

- Firstly, make absolutely sure that the seat is properly fitted in accordance with the manufacturer's instructions. If you're unsure about any of the fitting procedure, contact the manufacturer for advice.
- To hold a child securely, the harness must be reasonably tight. There should be just enough room to slide your flat hand under the strap. Children may be wearing bulky clothes one day, and thin clothes the next, so it is vital that the harness is adjusted before each journey to ensure a correct fit.
- If the seat is fitted using an adult's inertia reel seat belt, make sure that the seat is held firmly in position, and that the seat belt is securely locked to prevent it from loosening. Also, make sure that the seat belt buckle is not resting on the frame of the child seat. This is because the buckles are not designed to withstand the impact of a heavy child seat and, in an accident, could break open.

DRIVING ABROAD

Driving abroad can be very different to driving in the UK. Besides the obvious differences, like climate and driving on the right-hand side of the road, various unfamiliar laws may apply, and it's advisable to prepare yourself and your car as far as possible before travelling.

This Section provides you with a guide which will help you to prepare for driving abroad, and will help you to avoid some of the pitfalls waiting for the unwary.

It may be worthwhile considering hiring a car abroad, rather than driving your own car. In this case, make sure that the insurance cover arranged suits your requirements, and make sure that a damage waiver is included (otherwise you will have to pay for any damage to the hire car).

Insurance

● Motoring

Make sure that you have adequate insurance cover for your car and your luggage.

Check on the legal requirements for insurance in the country you're visiting, and always inform your insurance company that you're taking your car abroad – they will be able to advise you of any special requirements.

Most car insurance policies automatically give the minimum legally-required cover for driving in EC countries, but if you require the same level of cover as you have in the UK, you will normally need to obtain a 'Green Card' (an internationally-recognised certificate of insurance) from your insurance company.

● Medical

It's always advisable to take out medical insurance for the car occupants. Not all countries have a free emergency medical service, and you could find yourself with a large unexpected bill in the event of yourself or one of your passengers being taken ill, or being involved in an accident (in some countries you may even have to pay for an ambulance).

NHS form E111 (available by filling in a form at a Post Office) entitles you to receive

DRIVING ABROAD

the same health care that residents receive in EC countries, but this is by no means comprehensive, and additional insurance cover is usually advisable.

● Breakdown

Recovery and breakdown costs can be far higher abroad than in the UK.

Most of the national motoring organisations (such as the AA and RAC) will be able to provide insurance cover which could save you a lot of inconvenience and expense should you be unfortunate enough to break down.

Documents

Always carry your passport, driving licence, car registration document, and insurance certificate (including 'Green Card' and medical insurance, where applicable).

Make sure that all the documents are valid, and that the car's road fund licence (tax) and MOT don't run out while you're abroad.

Before travelling, check with the authorities in the country you're visiting, in case any special documents or permits are required. You may need a visa to visit some countries, and an International Driving Permit (available from the AA or RAC) is sometimes required.

Driving laws

Before driving abroad, make sure that you're familiar with the driving laws in the country you're visiting, as there may be some laws which don't apply in the UK, and the penalties for breaking the law may be severe.

Fit a 'GB' plate to the back of your car, and make sure that it's displayed all the time you're abroad.

In some countries, you're legally required to carry certain items of safety equipment. These can include a first-aid kit, a warning triangle, a fire extinguisher, a set of spare bulbs/fuses, etc.

Remember that if you're visiting a country where you have to drive on the right-hand side of the road, you'll need to fit headlight beam deflectors or shields to avoid dazzling other drivers, or alternatively, it may be possible to have the headlight beams adjusted.

Make sure that you're familiar with the speed limits, and note that in some countries there is an absolute ban on driving after drinking *any* alcohol.

Servicing

Service your car before setting off on your trip, to reduce the possibility of any unexpected breakdowns.

Pay particular attention to the condition of the battery, windscreen wipers and tyres (including the spare), noting that the tyre pressures will probably have to be increased from their normal setting if the car is to be fully loaded. If you're travelling to a cold country, check the condition of the cooling system and the strength of the antifreeze.

Before setting off on your journey, refer to *'Regular checks'* on page 77, and carry out all the checks described. Check that the jack and wheel brace are in place in the car (refer to *'Breakdowns'* on page 48), and that the jack works properly.

Spares

In addition to the items which must be carried by law in the country you're visiting, it's a good idea to carry a few spares which may be difficult to obtain should you need them abroad (one of the national motoring organisations should be able to advise you). For example, you may want to carry clutch and throttle cable repair kits, as right-hand-drive components can be difficult to find outside the UK.

If your car uses a special oil, it's a good idea to take a pack with you.

It's also a good idea to carry a tow rope and a set of jump leads to help you out in case of a breakdown.

Fuel

The type and quality of petrol available varies from country to country, and it's a good idea to check on the availability of the correct petrol type before travelling (again, one of the national motoring organisations will be able to advise you). This is especially important if your car has a catalytic converter, as in this case you must only use unleaded petrol.

Find out what petrol pump markings to look for to give you the correct type and grade of petrol for your car.

Security

A foreign car packed with luggage is an inviting prospect for criminals, so refer to *'Car*

DRIVING ABROAD

crime prevention' on page 71, and don't take any chances.

Route planning

It's always a good idea to plan your approximate route before travelling. A vast number of maps and guides are available, or the AA and RAC can provide you with directions to your destination for a modest charge.

Bear in mind that in some countries, you'll have to pay tolls to use certain roads, and this can add unexpectedly to the cost of travelling.

What to carry when driving abroad

The following list provides a guide to the items which it is compulsory to use or carry, or it is strongly recommended that you carry in your car when driving in various European countries.

In the following table **'C'** indicates **'Compulsory'** and **'R'** indicates **'Recommended'**.

COUNTRY	HEADLAMP DEFLECTORS	GB STICKER	SEAT BELTS	WARNING TRIANGLE	FIRE EXTINGUISHER	FIRST AID KIT	SPARE BULBS
Austria	C	C	C	C	-	C	-
Belgium	C	C	C	C	C	-	-
Bulgaria	C	C	C	C	C	C	-
Czechoslovakia	C	C	C	C	-	C	C
Denmark	C	C	C	R	-	-	-
Eire	-	C	C	-	-	-	-
Finland	C	C	C	C	-	-	-
France	C	C	C	C	-	-	R
Germany	C	C	C	C	C	C	R
Greece	C	C	C	C	C	C	-
Holland	C	C	C	C	-	-	C
Hungary	C	C	C	C	-	-	C
Italy	C	C	C	C	-	-	R
Luxembourg	C	C	C	R	-	-	-
Norway	C	C	C	R	-	-	R
Poland	C	C	C	C	-	-	R
Portugal	C	C	C	C	C	-	-
Spain	C	C	C	C	-	-	C
Sweden	C	C	C	R	-	-	-
Switzerland	C	C	C	C	-	-	-
United Kingdom	-	-	C	-	-	-	-
Yugoslavia	C	C	C	C	-	C	C

68 REDUCING THE COST OF MOTORING

VAUXHALL ASTRA & BELMONT

REDUCING THE COST OF MOTORING

ECONOMICAL DRIVING

Owning a car is always going to involve some expense, and the running costs can generally be divided into two main areas. The first concerns fixed costs which cannot be avoided, such as car tax and insurance (although obviously you can shop around for the best insurance quote). However, for most owners savings can be made in the second main area of expense, which covers fuel costs and servicing/maintenance bills.

It's surprisingly easy to reduce the amount of fuel used, simply by adapting your driving style to suit the prevailing conditions, and avoiding certain driving habits which tend to increase fuel consumption unnecessarily.

A large proportion of the average servicing bill is made up of garage labour time, money which can be saved by carrying out the work yourself. Refer to *'Servicing'* on page 89 for easy-to-follow instructions on how to carry out most servicing work. It's also worth bearing in mind that maintenance costs can be cut significantly by reducing unnecessary wear-and-tear on the car.

The following advice deals with the most significant causes of high fuel consumption and unnecessary wear-and-tear, and explains how to save money and reduce environmental pollution during everyday driving.

- **Don't warm the engine up with the car standing still** – Engine wear and pollution is at its highest when the engine is warming up, and the engine takes a long time to warm up when running at idle speed with the car stationary. To avoid excessive wear and pollution, drive off as soon as the engine starts, and don't use more 'revs' than necessary.

- **On cars with a manual choke, don't use the choke any longer than necessary** – Push the choke control fully in as soon as the engine will run smoothly without it. When the engine is running with the choke applied, extra fuel is being used and so the fuel consumption is increased, and more pollution is produced.

- **Avoid sudden full throttle acceleration** – Sudden acceleration increases fuel consumption, engine wear and pollution.

- **Don't drive at high engine speeds** – Minimum fuel consumption and pollution is achieved at low engine speeds and in the highest possible gear. Lower engine speed also means less noise and engine wear. For maximum economy, stay in as high a gear as possible for as long as possible, without making the engine labour.

- **Don't always drive at maximum speed** – Fuel consumption, pollution and noise increase rapidly at high speeds. A small reduction in speed (particularly during motorway driving) can significantly lower fuel consumption and pollution.

- **Look well ahead, and drive as smoothly as possible** – By looking well ahead you will be able to react to any change in road conditions in plenty of time, allowing you to brake and accelerate smoothly. Unnecessary or harsh acceleration and braking increases fuel consumption and pollution.

- **Where possible, avoid dense, slow-moving traffic** – In these conditions, more frequent braking, acceleration and gear changing are required, which increase fuel consumption and pollution.

- **Stop the engine during traffic hold-ups** – Obviously if your engine has stopped it doesn't use any fuel and there is no pollution.

VAUXHALL ASTRA & BELMONT

REDUCING THE COST OF MOTORING

- **Check the tyre pressures regularly** – Low tyre pressures increase the rolling resistance of the car, and therefore increase fuel consumption, as well as increasing tyre wear and causing handling problems.

- **Don't carry unnecessary luggage** – Weight has a significant effect on fuel consumption, especially in dense traffic where frequent acceleration is required.

- **Don't leave a roof rack fitted when not in use** – The extra air resistance increases fuel consumption.

- **Switch off any unnecessary electrical circuits as soon as possible** – Heated rear windows, foglights, heater blowers, etc, consume a considerable amount of electrical power. The engine must work harder to provide this power by driving the alternator, and the fuel consumption is therefore increased.

- **Check the fuel consumption regularly** – By doing this you will be able to notice any significant increase in fuel consumption, and any problem which may be causing it can be investigated before it develops into anything more serious.

- **Ensure that your car is serviced regularly** – This will ensure that the car operates as efficiently as possible, reducing fuel consumption and pollution.

- **If your car engine is suitable, use unleaded petrol** – This will reduce pollution. If in any doubt as to whether your car's engine is suitable for use with unleaded petrol, seek advice from the car's manufacturer, or from a recognised dealer.

VAUXHALL ASTRA & BELMONT

CAR CRIME PREVENTION

Crime against cars is a serious problem, and many owners suffer the consequences of theft or damage every year. The likelihood of your car becoming a target for criminals can be reduced by making it more difficult for your car to be broken into or stolen, and the following advice should prove helpful. Some of the points may seem obvious, but the majority of cars are broken into or stolen in a very short space of time with little force or effort required.

- **Always remove the ignition key** – even in a garage or driveway at home. Always make sure that the steering column lock is engaged after removing the key.
- **Always lock your car** – even in a garage or driveway at home. Where fitted, ensure that 'deadlocks' are engaged before leaving the car and, where applicable, don't forget to lock the fuel filler cover. Make sure that all the windows and the sunroof or folding roof, where applicable, are properly shut. If the car is in a garage, lock the garage. If your car is stolen or broken into while it's unlocked, your insurance company may not pay for the full value of loss or damage.
- **Never leave valuable items on display** – Even if you're only leaving your car for a few minutes, move anything that might be attractive to a thief (even a coat or briefcase) out of sight, and preferably take it with you or lock it safely in the boot. Don't leave valuable items (especially credit cards) in the glovebox. Don't leave your vehicle documents in the car (registration document, MOT certificate, insurance certificate, etc), as they could help a thief to sell it.
- **Park in a visible and (preferably) busy area** – This will deter thieves as they run a greater risk of being caught. If you're parking your car at night, try to leave it in a well-lit area.
- **Put your radio aerial down (where applicable) when parking** – Radio aerials can prove an attractive target for vandals.
- **Protect any in-car entertainment equipment** – The latest security-coded equipment won't work if someone tampers with it and disconnects it from the battery. Some equipment is specially designed so that you can remove the control panel or the entire unit, and take it with you when you leave the car.
- **Fit lockable wheel nuts (or bolts, as applicable)** – if your car is fitted with expensive alloy wheels. Alloy wheels are a favourite target for thieves.
- **Have your car windows etched with the registration number** – This will help to trace your car if it's stolen and the thieves try to change its identity. Other glass components such as sunroofs and headlamps can also be etched if desired. Many garages and specialists can provide this service, and it's possible to buy DIY glass etching kits from motor accessory shops if you prefer to tackle the job yourself.
- **Fit a vehicle immobiliser device** – Many different types are available, but the most common types consist of a substantial metal bar which can be locked in place between the steering wheel and the pedals to prevent the car from being driven.
- **Have an alarm fitted** – Many different types are available, and some are expensive, but they will deter thieves. Some alarms have built-in immobiliser devices to prevent the car from being driven. If you have an alarm fitted, remember to switch it on even if you're only leaving the car for a few minutes.

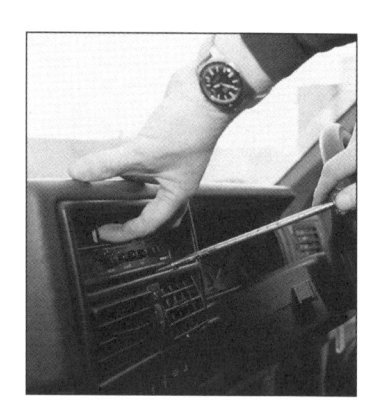

VAUXHALL ASTRA & BELMONT

72 SERVICE SPECIFICATIONS

VAUXHALL ASTRA & BELMONT

SERVICE SPECIFICATIONS

Recommended lubricants and fluids

Component/system	Lubricant/fluid type and specification	Duckhams recommendation
ENGINE (1)	Multigrade engine oil, viscosity range SAE 10W/40 to 20W/50, to API SF or SG	Duckhams QXR, QS, Hypergrade Plus or Hypergrade
COOLING SYSTEM (2)	55% clean water and 45% ethylene glycol-based antifreeze	Duckhams Universal Antifreeze and Summer Coolant
BRAKING SYSTEM (3)	Hydraulic brake fluid to SAE J1703 or DOT 4	Duckhams Universal Brake and Clutch Fluid
MANUAL GEARBOX (4)		
All 4-speed models	Gear oil, viscosity SAE 80, to API GL4	Duckhams Hypoid 80
All 5-speed models except Astra GTE	Gear oil, viscosity SAE 80, to API GL4	Duckhams Hypoid 75W/90S
Astra GTE	Special transmission fluid, GM part number 90 188 629	Duckhams Hypoid 75W/90S
AUTOMATIC TRANSMISSION (5)	Dexron II type ATF	Duckhams Uni-Matic
POWER STEERING (6)	Dexron II type ATF	Duckhams Uni-Matic

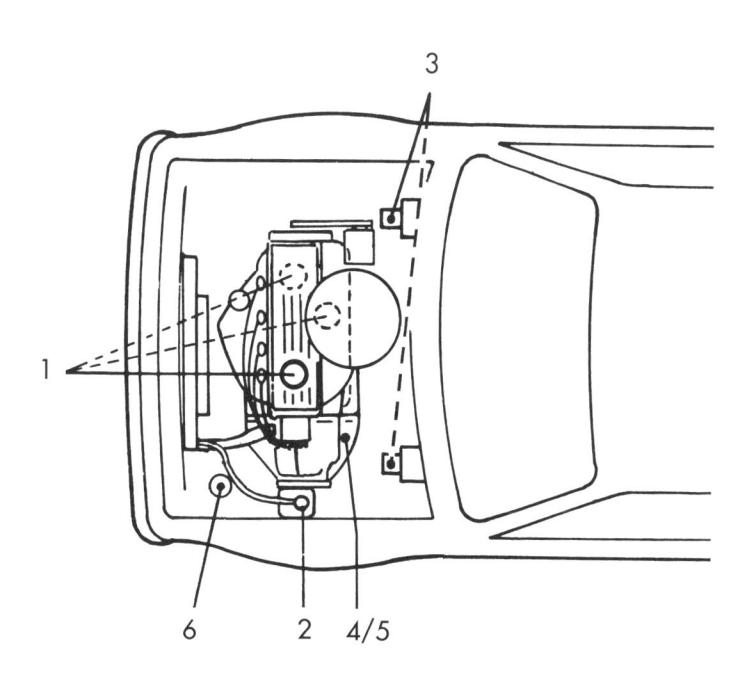

VAUXHALL ASTRA & BELMONT

SERVICE SPECIFICATIONS

Lubricant and fluid capacities [Litres (Pints)]

Fuel tank capacity

Hatchback/Saloon to end September 1985	**42** litres (9.3 gallons)
Hatchback/Saloon from October 1985	**52** litres (11.4 gallons)
Estate	**50** litres (11.9 gallons)

Engine oil
Quantity of oil required to bring level on dipstick from lower to upper mark:

1.2, 1.3 & 1.4 litre engines	**0.75** litre
All other engines	**1.0** litre

Engine oil capacity (including filter)

1.2 litre	**2.75** (4.8)
1.3 litre (up to chassis No C2 549 985)	**2.75** (4.8)
1.3 litre (from chassis No C2 549 985), and 1.4 litre	**3.00** (5.3)
1.6 & 1.8 litre to end September 1986	**3.25** (5.7)
1.6 litre from October 1986	**3.5** (6.2)
1.8 litre from October 1986, and 2.0 litre 8-valve	**4.0** (7.0)
2.0 litre 16-valve	**4.5** (7.9)

Coolant capacity (for coolant change)

1.2 litre 'Mark 1' models	**5.9** (10.4)
1.2 litre 'Mark 2' models	**5.7** (10.0)
1.3 litre 'Mark 1' models (manual)	**6.3** (11.1)
1.3 litre 'Mark 1' models (automatic)	**7.1** (12.5)
1.3 litre 'Mark 2' and Belmont models	**7.0** (12.3)
1.4 litre	**6.2** (10.9)
1.6 litre (manual) to end September 1988	**7.8** (13.7)
1.6 litre (manual) from October 1988	**6.4** (11.3)
1.6 litre (automatic) to end September 1988	**7.6** (13.4)
1.6 litre (automatic) from October 1988	**6.2** (10.9)
1.8 litre to end September 1988	**7.6** (13.4)
1.8 litre from October 1988	**6.9** (12.1)
2.0 litre	**6.9** (12.1)

VAUXHALL ASTRA & BELMONT

SERVICE SPECIFICATIONS

Contact breaker points
Points gap [mm (in)] **0.40** (0.016)

Spark plugs

Type	AC	Champion
1.2 litre to end March 1987	**R42 6FS**	**RL82YCC** or **RL82YC**
1.2 litre from April 1987	**CR42 6FS**	**RL82YCC** or **RL82YC**
1.3, 1.6 & 1.8 litre to end September 1986	**R42 XLS**	**RN7YCC** or **RN7YC**
1.3, 1.6 & 1.8 litre, from October 1986 to end March 1987	**R42 CXLS**	**RN7YCC** or **RN7YC**
1.3, 1.6 & 1.8 litre from April 1987, and 2.0 litre 8-valve	**CR42 CXLS**	**RN7YCC** or **RN7YC**
2.0 litre 16-valve	**(Bosch) 90297152**	**RC9MCC**

Electrode gap [**mm** (in)]

AC/Bosch	**0.7** to **0.8** (0.028 to 0.032)
Champion RL82YCC, RN7YCC & RC9MCC	**0.8** (0.032)
Champion RL82YC, RN7YC	**0.7** (0.028)

Clutch
Clutch pedal stroke [mm (in)] **138.0** (5.4)

VAUXHALL ASTRA & BELMONT

SERVICE SPECIFICATIONS

Tyre pressures - cold [bars (lbf/in^2)]

Note: *The following is intended as a guide only. Manufacturers frequently change tyre pressure recommendations, and it is suggested that a Vauxhall dealer is consulted for latest recommendations.*

	Front	Rear
Up to 3 occupants		
155 SR 13 (Hatchback/Saloon)	**1.9** (28)	**1.9** (28)
155 SR 13 (Estate)	**1.9** (28)	**2.2** (32)
155 TR 13 (Hatchback/Saloon)	**2.0** (29)	**1.8** (26)
165 R 13 (Hatchback/Saloon)	**1.8** (26)	**1.6** (23)
165 R 13 (Estate)	**1.8** (26)	**1.8** (26)
175/70 SR 13 (Hatchback)	**1.8** (26)	**1.8** (26)
175/70 SR 13 (Estate)	**1.8** (26)	**2.0** (29)
175/70 TR 13 (Hatchback/Saloon)	**2.0** (29)	**1.8** (26)
175/70 TR 13 (Estate)	**2.0** (29)	**2.0** (29)
175/65 SR 14 (Hatchback)	**1.9** (28)	**2.1** (31)
175/65 HR 14 (Hatchback/Saloon)	**2.1** (31)	**1.9** (28)
185/60 HR 14 (Hatchback/Convertible)	**2.0** (29)	**2.0** (29)
185/60 VR 14 (Hatchback)	**2.3** (33)	**2.1** (31)
185/65 VR 14 (Hatchback)	**2.3** (33)	**2.1** (31)
Fully laden		
155 SR 13 (Hatchback/Saloon)	**2.1** (31)	**2.4** (35)
155 SR 13 (Estate)	**2.2** (32)	**3.2** (46)
155 TR 13 (Hatchback/Saloon)	**2.1** (31)	**2.5** (36)
165 R 13 (Hatchback/Saloon)	**1.9** (28)	**2.1** (31)
165 R 13 (Estate)	**2.0** (29)	**2.8** (41)
175/70 SR 13 (Hatchback)	**2.0** (29)	**2.3** (33)
175/70 SR 13 (Estate)	**2.0** (29)	**2.8** (41)
175/70 TR 13 (Hatchback/Saloon)	**2.1** (31)	**2.3** (33)
175/70 TR 13 (Estate)	**2.1** (31)	**3.0** (44)
175/65 SR 14 (Hatchback)	**2.1** (31)	**2.4** (35)
175/65 HR 14 (Hatchback/Saloon)	**2.2** (32)	**2.4** (35)
185/60 HR 14 (Hatchback)	**2.2** (32)	**2.4** (35)
185/60 VR 14 (Hatchback)	**2.4** (35)	**2.6** (38)
185/65 VR 14 (Hatchback)	**2.4** (35)	**2.6** (38)

REGULAR CHECKS

To ensure that your car is reliable and safe to drive, there are one or two essential checks which are so simple that they're often ignored. These checks only take a few minutes, and could save you a lot of inconvenience and expense. It's a good idea to carry out these checks once a week, and certainly before you start off on a long journey. This Section explains how to carry out the checks, and what to do if things aren't quite as they should be.

Whenever you're carrying out checks or servicing jobs, safety must always be the first consideration, and you should bear in mind the advice given in the *'Safety first!'* on page 98 before proceeding.

CHECKS

The following checks should be carried out regularly. The checks are explained in more detail in the following pages.
- Check the oil level
- Check the coolant level
- Check the brake fluid level
- Check the tyres
- Check the washer fluid level
- Check the battery electrolyte level (where applicable)
- Check the wipers and washers
- Check the lights and horn
- Check for fluid leaks

ITEMS REQUIRED WHEN CARRYING OUT CHECKS

When carrying out the regular checks, it's a good idea to have the following items close-to-hand. You won't always need all the items, but it's as well to have them available just in case.

- Small quantity of clean rag ● 1.0 litre pack of engine oil (of the correct type – refer to **'Service specifications'** on page 73) ● Small quantity of coolant solution (made up from approximately 55% clean water and 45% antifreeze) ● Small container of brake fluid (must be an airtight container) ● Tyre pressure gauge and foot pump ● Small screwdriver (for removing stones from tyres) ● Washer fluid additive ● Small quantity of distilled water – for models fitted with batteries which require topping-up ● Pin (for adjusting washer nozzles)

There are several different engine sizes and types fitted to the Astra/Belmont range of models. Refer to *'Servicing'* on page 101 for details of how to distinguish one engine from another.

REGULAR CHECKS

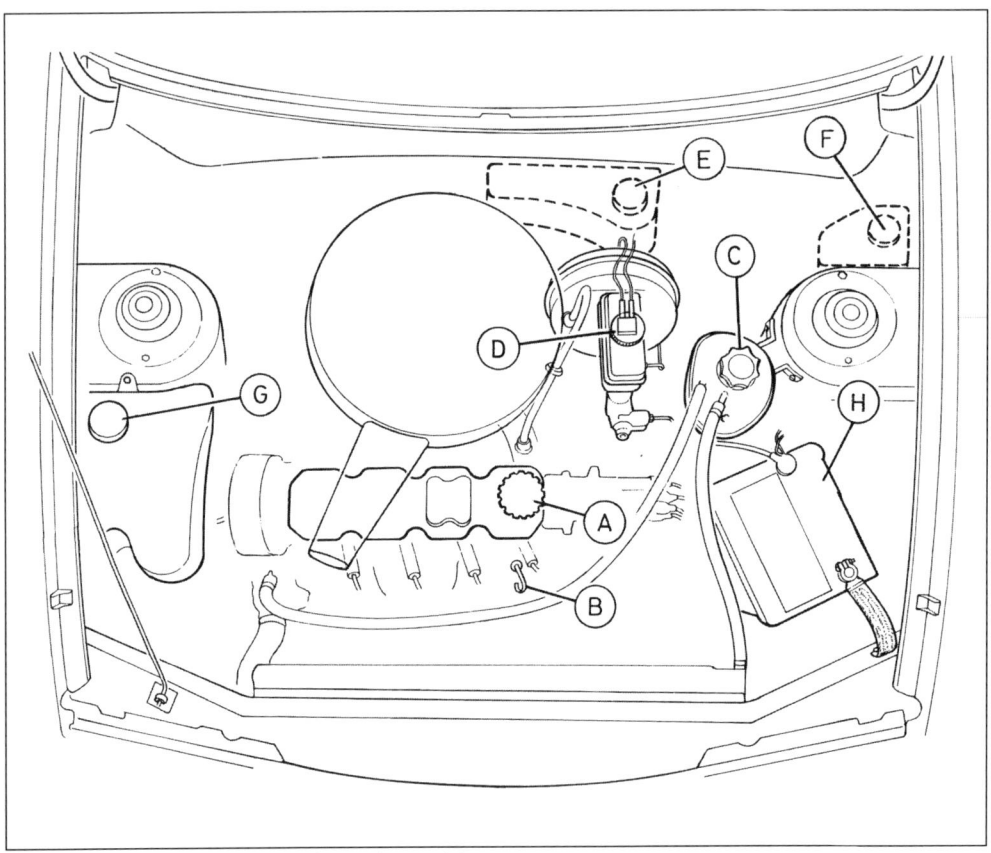

▲ *Typical engine compartment layout for 'Mark 1' models*
  A *Engine oil filler cap*
  B *Engine oil dipstick*
  C *Coolant expansion tank*
  D *Brake fluid reservoir*
  E *Windscreen washer fluid reservoir (most models)*
  F *Auxiliary washer fluid reservoir (models with headlamp washers)*
  G *Main washer fluid reservoir (models with headlamp washers)*
  H *Battery*

REGULAR CHECKS 79

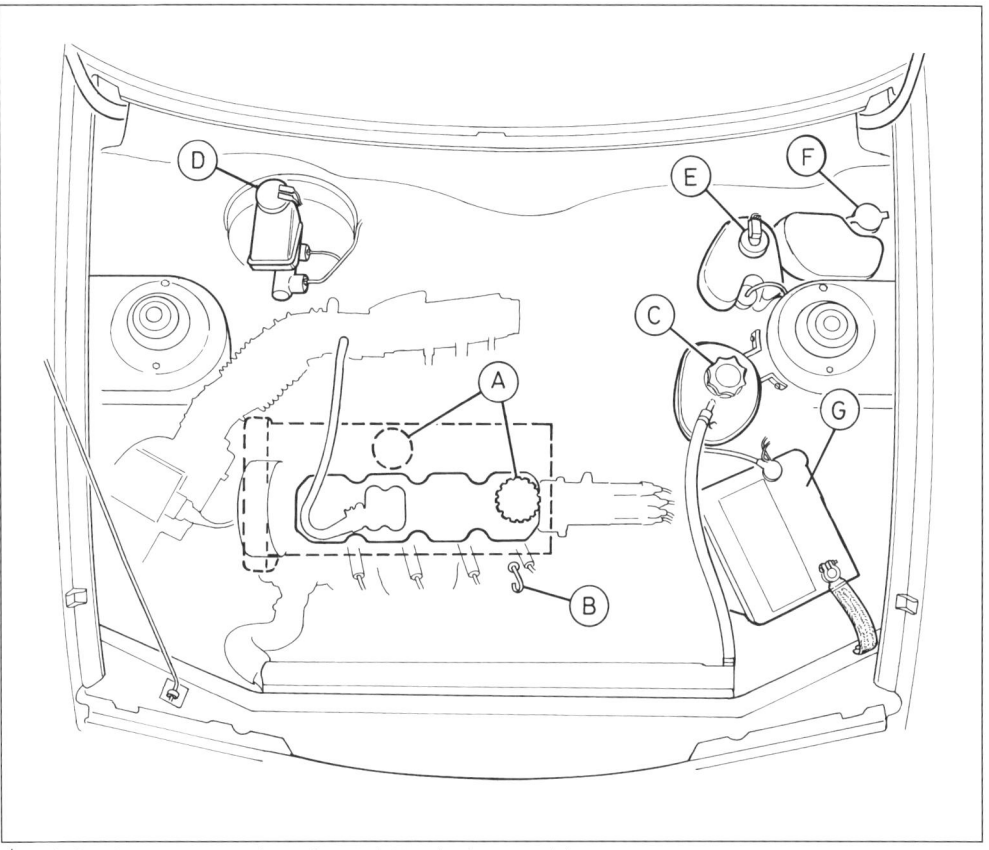

▲ *Typical engine compartment layout for 'Mark 2' and Belmont models*
  A *Engine oil filler cap (alternative location also shown)*
  B *Engine oil dipstick*
  C *Coolant expansion tank*
  D *Brake fluid reservoir*
  E *Windscreen washer fluid reservoir*
  F *Headlamp washer reservoir filler*
  G *Battery*

VAUXHALL ASTRA & BELMONT

REGULAR CHECKS

Checking oil level

To check the oil level, the car should be standing on level ground, and the engine should have been stopped for at least a few minutes. Open the bonnet, and look for the dipstick, which is located on the front of the engine. On the overhead valve 1.2 litre engine it is located next to the distributor, but on the overhead camshaft 1.3, 1.6 & 1.8 litre engines it is located in a tube.

Pull out the dipstick and wipe it clean (with a clean non-fluffy cloth), slowly push it fully back into its location, and pull it out again. Note the oil level, which should never be allowed to drop below the bottom mark on the dipstick. Where the dipstick incorporates an oil level sensor, it will have wires attached to it, and should be treated with care as it is fragile.

If the oil needs topping-up, unscrew and remove the oil filler cap and top-up until the dipstick reading is on the upper level mark.

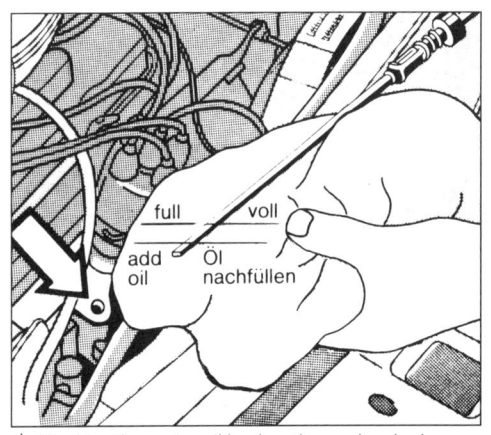

▲ Topping-up the engine oil level

Note that the amount of oil required to raise the level on the dipstick from the lower to upper marks is 0.75 litre (1.3 pint) on 1.2 & 1.3 litre engines, and 1.0 litre (1.8 pint) on all other engines. Always try to use the same type and make of oil, and take care not to overfill. Mop up any oil which might have been spilt, and make sure that the oil filler cap is correctly refitted.

If the oil needs to be topped-up regularly, check for leaks, and if necessary seek advice.

▲ Checking the engine oil level on the overhead valve 1.2 litre engine

Checking coolant level

The coolant level can be checked visually by checking the level in the expansion tank when the engine is **cold**. The level should be a little above the KALT (cold) mark.

If topping-up is necessary, first the cooling system pressure cap must be removed. Topping-up should always be done with the engine cold, but if it does prove necessary to remove the cooling system pressure cap with the engine hot for any reason, take care to avoid scalding. Place a thick rag over the cap, and loosen the cap slowly in stages to gradually release the pressure in the system.

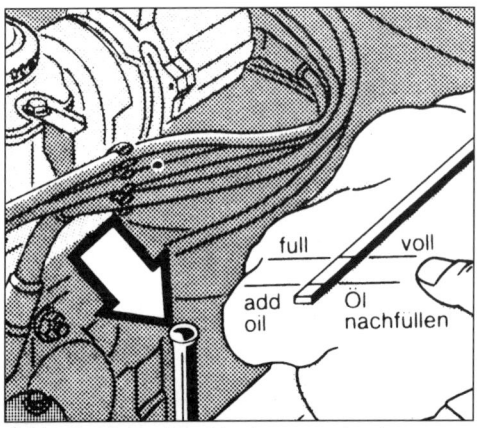

▲ Checking the engine oil level on an overhead camshaft engine

VAUXHALL ASTRA & BELMONT

REGULAR CHECKS

▲ *Expansion tank showing KALT (cold) marking*

A screw-type pressure cap is fitted to the expansion tank. Slowly unscrew the cap until all the pressure in the system is released, then remove the cap.

Topping-up should always be done with a mixture of water and antifreeze to the same strength as the mixture already in the system (in this case, 55% clean water to 45% antifreeze). It's important to note that because of the different types of metals used in the engine, it's vital to use antifreeze with suitable anti-corrosion additives all year round.

Although water alone can be used for topping-up, it's unwise to make it a habit, as the strength of the antifreeze in the main system will gradually be diluted. **Never** use water alone to fill the whole system. After topping-up, refit and tighten the filler cap.

Normally, topping-up will rarely be required, and if the need for regular topping-up arises, it will probably be due to a leak somewhere in the system. Leaks are most likely to occur from the radiator or the various hoses. If no leaks can be found, it's possible that there's an internal fault in the engine, such as a crack in the cylinder head, or a blown cylinder head gasket, but in this case it's best to seek specialist advice.

Checking brake fluid level

Note: *Refer to* **'Safety first'** *on page 98 for special precautions which should be taken when handling brake fluid.*

Although a low brake fluid level warning light is fitted, the fluid level should be checked visually whenever the oil and coolant levels are checked. The level should be up to the 'MAX' mark on the side of the reservoir, but it is quite normal for the level to fall slightly as the brake friction material wears.

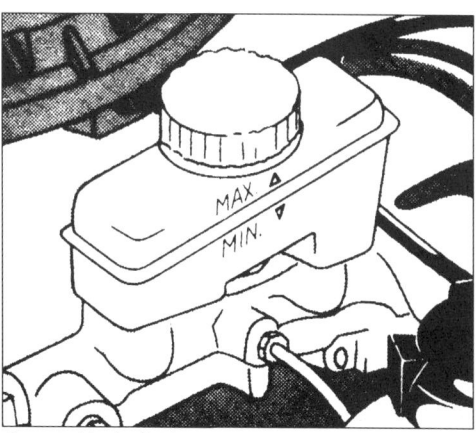

▲ *Typical brake fluid reservoir and markings*

The fluid level must **never** be allowed to drop below the 'MIN' mark.

If topping-up is required, always use the correct type of fluid, which should always be stored in a full, airtight container. Don't top-up using fluid which has been stored in a partly-full container, as it will have absorbed moisture from the air, which can dangerously reduce its performance.

▲ *Topping-up the coolant level through the expansion tank*

VAUXHALL ASTRA & BELMONT

Topping-up should hardly ever be required, unless there's a leak somewhere in the hydraulic system. If a leak is suspected, the car should not be driven until the braking system has been thoroughly checked. **Never** take any risks where brakes are concerned.

Checking tyres

It's extremely important to carry out regular checks on the tyres, to make sure that the pressures are correct, and that the tyres are not damaged. The tyres are the only part of the car in contact with the road, so their condition will affect the steering and general handling of the car, and therefore its safety.

To check tyre pressures accurately, the tyres must be cold, which means that the car must not have been driven recently. Note that it can make a noticeable difference to the tyre pressures if the car has been standing out in the sun. This can be very noticeable when one side of the car is in the sun, and the other side is in shadow.

Note that the recommended tyre pressures vary depending on the size of tyres fitted – refer to *'Service specifications'* on page 76. The size of the tyre (eg 155 SR 13) is clearly marked on the tyre sidewall, although the style of the tyre size marking may vary depending on the tyre manufacturer – consult a Vauxhall dealer or a tyre specialist for further details of tyre size markings, and the latest pressure recommendations.

If the tyres are being checked after the car has been driven, for example when filling up with petrol during a journey, the pressures are bound to be higher than specified. It's best to simply check that the pressures in the two front tyres are *equal*, and similarly for the two back tyres (remember that the specified pressure for the rear tyres may be different to the front). **Never** let air out of the tyres of a car which has

Shoulder wear

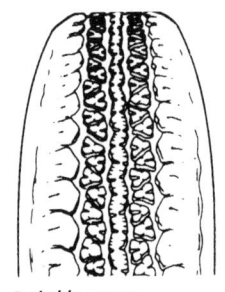

Probable cause:
Underinflation (wear on both sides)
Action: Check and adjust pressure

Probable cause:
Incorrect wheel camber (wear on one side)
Action: Repair or renew suspension parts

Probable cause:
Hard cornering
Action: Reduce speed

Centre wear

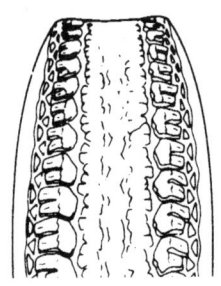

Probable cause:
Overinflation
Action: Measure and adjust pressure

Uneven wear

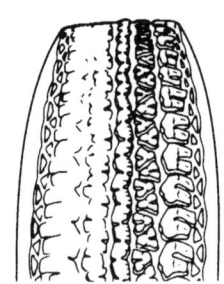

Probable cause:
Incorrect camber or castor
Action: Repair or renew suspension parts

Probable cause:
Malfunctioning suspension
Action: Repair or renew suspension parts

Probable cause
Unbalanced wheel
Action: Balance tyres

Probable cause:
Out-of-round brake disc/drum
Action: Machine or renew disc/drum

Toe wear

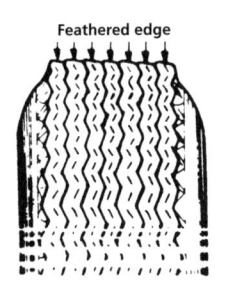

Feathered edge

Probable cause:
Incorrect toe setting
Action: Adjust front wheel alignment

▲ *Tyre wear patterns and causes*

REGULAR CHECKS 83

recently been driven, to bring the pressures down to that specified (unless the tyre pressures had been increased for a fully-loaded car, and the load has just been reduced).

When checking the tyre pressures, don't forget to check the spare!

With the tyres properly inflated, run your fingers around the edge of the tyre to check for any cuts and bulges in the tyre walls. Ideally, the car should be jacked up to visually check the tyres all round but, alternatively, the car can be rolled backwards and forwards to enable you to check all round the tread. Check the tread for cuts, and remove any debris, such as small stones and broken glass, using a screwdriver or similar tool. Any large items such as nails which have penetrated the surface of the tyre should be left in the tread (to identify the location of the damage) until the tyre has been repaired. Refer to 'Breakdowns' on page 48 for details of how to fit the spare wheel.

Also check the tyre treads for wear. Uneven wear across one particular tyre may be due to a fault with the suspension or steering.

By law, the tread depth must be at least 1.6 mm (0.06 in), throughout a continuous band comprising the central three-quarters of the width of the tyre tread, around the full circumference of the tyre.

If you find that any of the tyres is excessively worn or damaged, obtain a new tyre as soon as possible. It's a good idea to try and stick to one type of tyre if possible, rather than mixing several different makes on the car. Generally, you'll find that it's much cheaper to buy tyres from a tyre specialist, rather than an ordinary garage. Make sure that when you buy a new tyre you have the wheel balanced, otherwise you might find that the new tyre causes vibration when driving.

Checking washer fluid level

'MARK 1' MODELS

On 'Mark 1' models not fitted with headlamp washers, the windscreen washer fluid reservoir is located on the bulkhead in the engine compartment, behind the air cleaner.

On non-GTE models, if a headlamp washer system is fitted, the main reservoir is located in

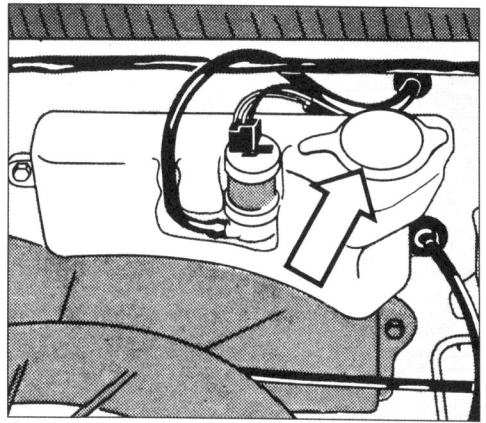

▲ *Windscreen washer fluid reservoir location on most 'Mark 1' models*

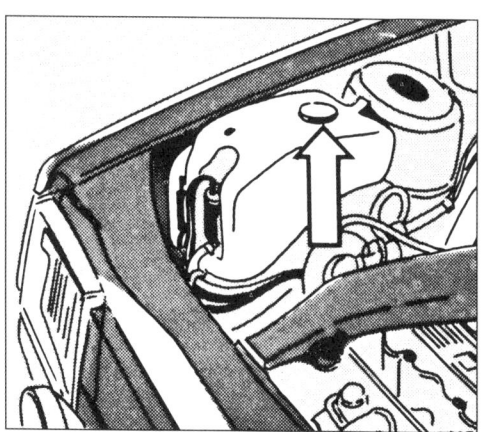

▲ *Main washer fluid reservoir location on non-GTE models fitted with a headlamp washer*

the right-hand front corner of the engine compartment. A smaller reservoir may also be found in the right-hand rear corner of the engine bay.

On GTE models with a headlamp washer, the reservoir is located in the left-hand wheel arch, and is filled through a filler neck in front of the battery.

The tailgate washer fluid reservoir is located behind a cover in the left-hand rear corner of the luggage compartment on Hatchback models, and beneath the spare wheel cover on the right-hand side on Estate models.

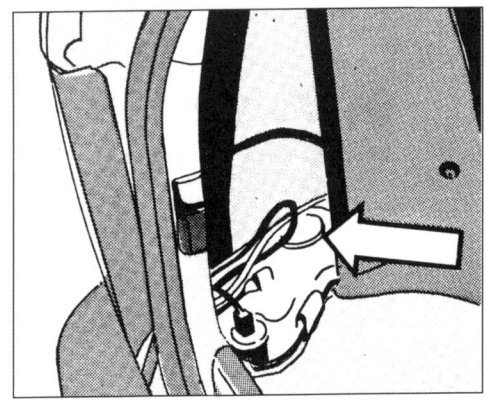

REGULAR CHECKS

▲ *Tailgate washer fluid reservoir location on 'Mark 1' Hatchback models*

▲ *Windscreen washer fluid reservoir location on 'Mark 2' and Belmont models*

▲ *Tailgate washer fluid reservoir location on 'Mark 1' Estate models*

▲ *Headlamp washer fluid reservoir filler location on 'Mark 2' and Belmont models*

'MARK 2' AND BELMONT MODELS

On 'Mark 2' and Belmont models, the windscreen washer fluid reservoir is located in the left-hand rear corner of the engine compartment.

On models with headlamp washers, the reservoir has two filler caps – one for the windscreen washer and the other for the headlamp washer.

ALL MODELS

If topping-up is necessary, pull the cap from the reservoir, and top-up as necessary with clean water. A suitable washer fluid additive will keep the glass free from smears, and will prevent the fluid freezing in Winter.

Checking battery electrolyte level

Except on early models, the battery fitted as original equipment is of the maintenance-free type, which is 'sealed for life', and does not require topping-up with distilled water. A maintenance-free battery is usually clearly marked (typically 'Freedom Battery'), and there will be no removable covers to enable topping-up.

On early models with a standard type battery, or on later models where the original battery has been replaced by one of this type (or of a 'low maintenance' type) from another source, the electrolyte level should be checked as follows.

VAUXHALL ASTRA & BELMONT

REGULAR CHECKS 85

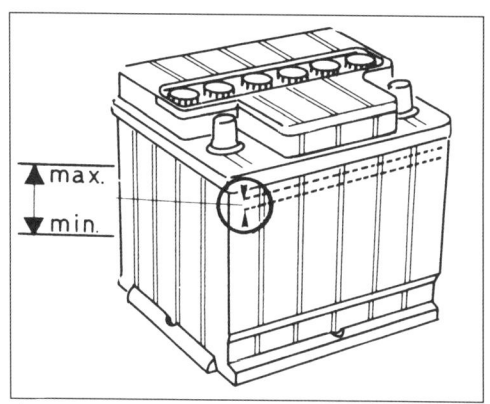

▲ *Electrolyte level markings on the standard battery with translucent case*

The battery is located in the front left-hand corner of the engine compartment (when viewed from the driver's seat). Remove the caps or the cover from the top of the battery, and check the level of the electrolyte fluid (with some types of battery, you can see the fluid level through the battery case).

The level should be just above the tops of the plates inside the battery, or in some cases up to a mark on the battery case.

If necessary, add distilled water (**not** ordinary tap water) to each cell, to bring the level just above the tops of the battery plates or up to the marks, as applicable.

With some types of battery, distilled water is added to a trough in the top of the battery until all the filling slots are full, and the bottom of the trough is just covered. On completion, refit the caps or cover, and carefully wipe up any drops of water that were spilt.

Topping-up should hardly ever be required, and the need for frequent topping-up indicates that the battery is being overcharged, due to a fault in the charging circuit – seek advice from someone suitably qualified if this is the case.

Checking wipers and washers

Check the wipers (front and rear, where applicable) and all washers to make sure that they're working properly. Don't allow the wipers to work on dry glass for too long, as it will strain the motor and wear out the wiper blades. If necessary, the washer nozzles can be adjusted using a pin inserted into the end of the nozzle. If you're adjusting the windscreen washer nozzles, remember to aim them fairly high on the windscreen, as the airflow will usually deflect the spray down when the car is moving.

Over a period of time, the wiping action of the wiper blades will deteriorate, causing smearing of the glass. The rubber may also crack, particularly at the edges of the blades. When this happens, the blades must be renewed. This is a straightforward job, and is accomplished as follows:

Pull the wiper arm away from the glass, against the spring pressure, until the arm clicks into position. Swivel the blade on the arm, then depress the catch on the U-shaped retainer, and slide the blade down then up from the wiper arm.

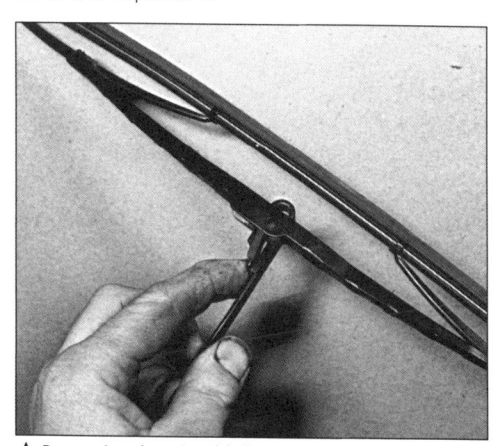

▲ *Topping-up the battery electrolyte level*

▲ *Depressing the wiper blade retaining catch*

VAUXHALL ASTRA & BELMONT

REGULAR CHECKS

On 'Mark 1' Estate models, to remove the tailgate wiper blade, prise off the blade pivot cover at the end of the wiper arm, and remove the blade retaining clip.

Disconnect the blade pivot from the arm, and then unscrew and remove the screw securing the blade assembly to the link, to free the blade assembly.

The blade is normally renewed complete, but the rubber insert can be renewed separately – however, this is a very fiddly job.

Checking lights and horn

Switch on all the lights in turn, and check that they're working. Don't forget to check the direction indicators and the brake lights (the

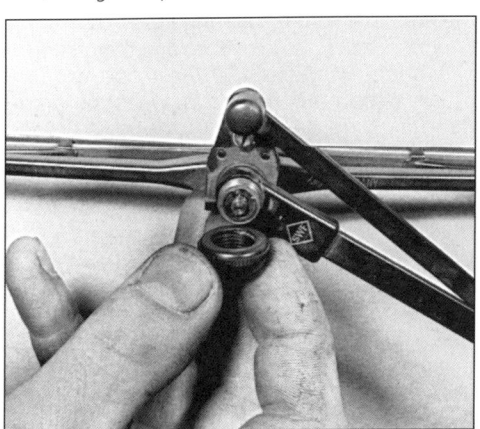

▲ Removing the wiper blade from the arm

▲ Removing the rear wiper arm pivot cover on 'Mark 1' Estate models

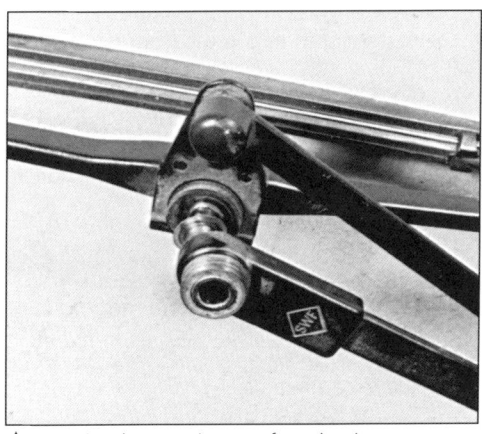

▲ Removing the rear wiper arm from the pivot on 'Mark 1' Estate models

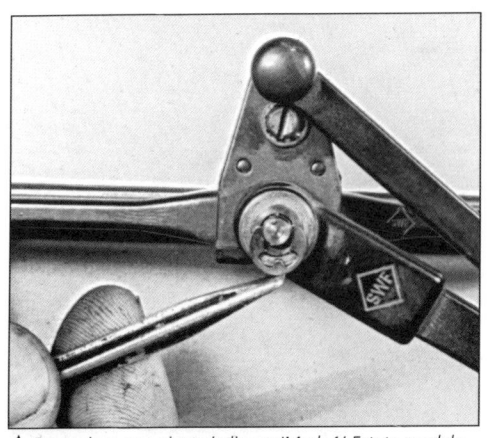

▲ Rear wiper arm pivot circlip on 'Mark 1' Estate models

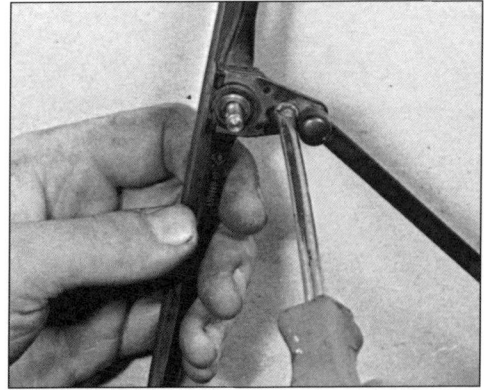

▲ Removing the rear wiper arm link screw on 'Mark 1' Estate models

VAUXHALL ASTRA & BELMONT

REGULAR CHECKS

ignition must be switched on to check these), either with the help of an assistant, or by looking for the reflection in a suitable window or door etc.

Check that the direction indicators work with the brake lights on, and that the brake lights work with the tail lights on. Some faults may cause the various rear lights to interact, which can be dangerous, as it may confuse or distract following drivers.

Check the operation of the horn. A short 'blast' should be enough to prove that it at least works, but from time to time check its operation over several seconds.

If any of the bulbs need renewing, refer to *'Bulb, fuse and relay renewal'* on page 131. Any other problems are likely to be caused by a faulty switch, or loose or corroded connections (especially earth connections), and it's best to seek advice to solve these.

Checking for fluid leaks

Check the ground where the car is normally parked for any stains or spots of fluid, which may have been caused by fluid leaking from the car.

Open the bonnet, and make a quick check of all the hoses and pipes, and the surfaces of all the components in the engine compartment. If there's any sign of leaking fluid, try to find the source of the leak, and seek advice if necessary. It can be difficult to identify leaking fluids, but if there's obviously a major leak, or if you suspect even a slight brake fluid or petrol leak, don't drive the car until the problem has been investigated by someone suitably qualified.

VAUXHALL ASTRA & BELMONT

SERVICING

VAUXHALL ASTRA & BELMONT

SERVICING

Regular servicing will ensure that your car is reliable and safe to drive, and could save you a lot of money in the long run. Many of the unexpected expenses which can crop up if things go wrong with your car can be avoided by carrying out regular servicing.

A significant proportion of the average garage servicing bill (well over 50% in many cases) is made up of labour costs, so obviously a lot of money can be saved by carrying out the work yourself. You'll find that most of the servicing jobs are very straightforward, and you don't need to be a mechanical genius to 'have a go' yourself. Don't be put off when you look under your bonnet; there are surprisingly few items which require frequent attention, and most of those which do are easily accessible without the need for anything more than basic tools. You'll discover that DIY servicing can be very rewarding, and will help you to understand what makes your car 'tick'.

The following chart lists all the servicing tasks recommended by the car manufacturer, and details of how to carry out the necessary work can be found in the subsequent pages. A few of the jobs require more extensive knowledge, or the use of special tools, and detailed explanation is beyond the scope of this Handbook. These more complicated tasks are identified on the chart, and details can be found in our Owners Workshop Manual for your particular model. Even if you decide to have the work done by a garage, the chart will enable you to check that the necessary work has been carried out.

Whenever you're carrying out servicing, safety must always be the first consideration, and you should read through the 'Safety first!' notes on page 98 before proceeding any further.

SERVICE SCHEDULE

ALL MODELS TO END SEPTEMBER 1982

ENGINE COMPARTMENT
Check all components for fluid leaks, corrosion or deterioration

ENGINE
Check engine oil level

Renew engine oil & filter

Check valve clearances (1.2 litre engine only)

COOLING SYSTEM
Check coolant level

Check antifreeze concentration

FUEL AND EXHAUST SYSTEMS
Clean fuel pump filter screen (early carburettor models)

Lubricate throttle/choke linkages

Renew air filter cleaner element

Clean fuel inlet filter gauze (where fitted)

Check exhaust system

Check idle speed/mixture settings

Check exhaust emissions

Renew fuel filter (fuel injection models only)

IGNITION SYSTEM
Clean contact breaker points (1.2 & 1.3 litre only)

Renew contact breaker points (1.2 & 1.3 litre only)

Check/adjust ignition timing (where contact breaker points fitted)

Check spark plugs

Renew spark plugs

CLUTCH
Check clutch pedal adjustment

MANUAL GEARBOX
Check gearbox oil level

AUTOMATIC TRANSMISSION
Check fluid level

Renew fluid

Note: *In addition to the above, also carry out the following task:*
Every 2 years, regardless of mileage, *renew the coolant (refer to* **'Service tasks'** *on page 123).*

VAUXHALL ASTRA & BELMONT

SERVICE SCHEDULE 91

SERVICE SCHEDULE 1 — ALL MODELS TO END SEPTEMBER 1982

Regular Weekly	6000 miles (10 000 km) 6 months	12 000 miles (20 000 km) 1 year	24 000 miles (40 000 km) 2 years
■	■	■	■
■	■	■	■
	■	■	■
	☐	☐	☐
■	■	■	■
	■	■	■
	■	■	■
	■	■	■
		■	■
		■	■
	☐	☐	☐
	☐	☐	☐
			☐
	■		
	■	■	■
	■		
		■	■
		■	■
	■	■	■
	■	■	■
			☐

Items marked ☐ are considered to be beyond the scope of this Handbook, and details can be found in the Owners Workshop Manual for your car (OWM 635).

VAUXHALL ASTRA & BELMONT

SERVICE SCHEDULE

ALL MODELS TO END SEPTEMBER 1982

BRAKING SYSTEM
Check brake fluid level
Check handbrake adjustment & linkages
Check front (and rear where applicable) disc pads
Check brake pressure-proportioning valve – where fitted (refer to note at end of table)
Check rear brakes/drums
Adjust rear brakes

SUSPENSION AND STEERING
Check tyres for condition & pressure
Check tightness of roadwheel bolts
Check power steering fluid level
Check front wheel alignment
Check suspension/steering balljoints

INTERIOR AND BODYWORK
Lubricate hinges/door locks/bonnet release mechanism

ELECTRICAL SYSTEM
Check washer fluid level
Check battery electrolyte level (where applicable)
Check wipers/washers
Check operation of lights & electrical equipment
Check drivebelts
Check headlamp beam alignment

CAR UNDERSIDE
Check all components for fluid leaks, corrosion or deterioration

ROAD TEST
Check instruments/electrical equipment
Check steering, suspension & general handling
Check engine, clutch (where applicable), gearbox & driveshafts
Check braking system

Note: *For a comprehensive check of the brake pressure-proportioning valve, refer to a Vauxhall dealer. In addition to the above, also carry out the following task:*
Every 12 months, regardless of mileage, *renew the brake hydraulic fluid (refer to the Owners Workshop Manual).*

VAUXHALL ASTRA & BELMONT

SERVICE SCHEDULE

93

Regular Weekly	6000 miles (10 000 km) 6 months	12 000 miles (20 000 km) 1 year	24 000 miles (40 000 km) 2 years
■	■	■	■
		■	■
	☐	☐	☐
	☐	☐	☐
		☐	☐
		☐	☐
■	■	■	■
	■	■	■
		■	■
	☐	☐	☐
		☐	☐
	■	■	■
■	■	■	■
■	■	■	■
■	■	■	■
■	■	■	■
		■	■
		☐	☐
	■	■	■
	■	■	■
	■	■	■
	■	■	■
	■	■	■

SERVICE SCHEDULE 2 — ALL MODELS TO END SEPTEMBER 1982

Items marked ☐ are considered to be beyond the scope of this Handbook, and details can be found in the Owners Workshop Manual for your car (OWM 635).

VAUXHALL ASTRA & BELMONT

SERVICE SCHEDULE

ALL MODELS FROM OCTOBER 1982

ENGINE COMPARTMENT
Check all components for fluid leaks, corrosion or deterioration

ENGINE
Check engine oil level

Renew engine oil & filter (every 6 months on models up to end September 1986)

Check valve clearances (1.2 litre engine only)

Check timing belt (all models except 2.0 litre 16-valve)

COOLING SYSTEM
Check coolant level

Check antifreeze concentration

FUEL AND EXHAUST SYSTEMS
Clean fuel pump filter screen (early carburettor models)

Check exhaust system

Lubricate throttle/choke linkages

Renew air filter cleaner element

Clean fuel inlet filter gauze (where fitted)

Check idle speed/mixture settings (except fuel injection models, October 1986 on)

Check exhaust emissions

Renew fuel filter (fuel injection models only)

IGNITION SYSTEM
Renew contact breaker points (1.2 litre engine only)

Check/adjust ignition timing (1.2 litre engine only)

Check spark plugs

Renew spark plugs

CLUTCH
Check clutch pedal adjustment

MANUAL GEARBOX
Check gearbox oil level (all models to end July 1991)

Check gearbox oil level (August 1991 on)

AUTOMATIC TRANSMISSION
Check fluid level (all models to end July 1991)

Check fluid level (August 1991 on)

Renew fluid

In addition to the above, also carry out the following tasks:
Every 2 years, regardless of mileage, *renew the coolant (refer to* **'Service tasks'** *on page 123).*

VAUXHALL ASTRA & BELMONT

SERVICE SCHEDULE

SERVICE SCHEDULE 1 — ALL MODELS FROM OCTOBER 1982

Regular / Weekly	9000 miles (15 000 km) 1 year	18 000 miles (30 000 km) 2 years	36 000 miles (60 000 km) 4 years
■	■	■	■
■	■	■	■
	■	■	■
	☐	☐	☐
			☐
■	■	■	■
	■	■	■
	■	■	■
	■	■	■
	■	■	■
		■	
	☐	☐	☐
	☐	☐	☐
		☐	☐
	■	■	■
	■	■	■
	■		
		■	■
	■	■	■
		■	■
	■	■	■
		■	
			☐

Items marked ☐ are considered to be beyond the scope of this Handbook, and details can be found in the Owners Workshop Manual for your car (OWM 635 for 'Mark 1' models, and OWM 1136 for 'Mark 2' and Belmont models).

VAUXHALL ASTRA & BELMONT

SERVICE SCHEDULE

ALL MODELS FROM OCTOBER 1982

BRAKING SYSTEM
Check brake fluid level
Check handbrake adjustment & linkages
Check front/rear disc pads
Check brake pressure-regulating valve (see note at end of table)
Check rear brakes/drums
Adjust rear brakes (all models to end September 1983)

SUSPENSION AND STEERING
Check tyres for condition & pressure
Check tightness of roadwheel bolts
Check power steering fluid level
Check front wheel alignment
Check suspension/steering balljoints

INTERIOR AND BODYWORK
Lubricate hinges/door locks/bonnet release mechanism

ELECTRICAL SYSTEM
Check washer fluid level
Check battery electrolyte level (where applicable)
Check wipers/washers
Check operation of all lights & electrical equipment
Check drivebelts
Check headlamp beam alignment

CAR UNDERSIDE
Check all components for fluid leaks, corrosion or deterioration

ROAD TEST
Check instruments/electrical equipment
Check steering, suspension & general handling
Check engine, clutch (where applicable), gearbox & driveshafts
Check braking system

Note: *For a comprehensive check of the brake pressure-regulating valve, refer to a Vauxhall dealer. In addition to the above, also carry out the following tasks:*
Every 9000 miles (15 000 km) or 12 months, *check the catalytic converter (where fitted) – refer to a Vauxhall dealer.*
Every 12 months, regardless of mileage, *renew the brake fluid (refer to the Owners Workshop Manual).*
On 2.0 litre 16-valve engines only, *the timing belt should be renewed every 8 years or 74 000 miles (120 000 km) – refer to the Owners Workshop Manual.*

VAUXHALL ASTRA & BELMONT

SERVICE SCHEDULE

SERVICE SCHEDULE 2 — ALL MODELS FROM OCTOBER 1982

Regular / Weekly	9000 miles (15 000 km) / 1 year	18 000 miles (30 000 km) / 2 years	36 000 miles (60 000 km) / 4 years
■	■	■	■
		■	■
	☐	☐	☐
	☐	☐	☐
		☐	☐
		☐	☐
■	■	■	■
	■	■	■
		■	■
		☐	☐
		☐	☐
	■	■	■
■	■	■	■
■	■	■	■
■	■	■	■
■	■	■	■
	■	■	■
	☐	☐	☐
	■	■	■
	■	■	■
	■	■	■
	■	■	■
	■	■	■

Items marked ☐ are considered to be beyond the scope of this Handbook, and details can be found in the Owners Workshop Manual for your car (OWM 635 for 'Mark 1' models, and OWM 1136 for 'Mark 2' and Belmont models).

VAUXHALL ASTRA & BELMONT

SERVICING

SAFETY FIRST!

Professional motor mechanics are trained in safe working procedures. No matter how enthusiastic you may be about getting on with the job you have planned, take time to read this Section. Don't risk an injury by failing to follow the simple safety rules explained here. The following is a guide to encourage you to be 'safety conscious' as you work on your car.

Essential points

ALWAYS use a safe system of supporting your car when working underneath it. The car's own jack (or a single hydraulic jack) is never sufficient. Once you've raised the car, you should use a safe means of holding it up. Axle stands, ramps or substantial wooden blocks are good; concrete blocks and bricks may crack or disintegrate, and should not be used. Make sure that you locate the supports where you know the car won't collapse, and where they can't slip.

ALWAYS loosen wheel nuts and other 'high torque' nuts or bolts (as applicable for the task to be carried out) before jacking up your car, or it may slip and fall.

ALWAYS make sure that your car is in **Neutral** or in **Park** (in the case of automatic transmission), and that the handbrake is securely on before trying to start the engine.

UNLESS the engine is cold, **ALWAYS** cover the cooling system's pressure (expansion tank) cap with several thicknesses of cloth before trying to remove it. Release the pressure slowly or the coolant may escape suddenly and scald you.

ALWAYS wait until the engine has cooled down before draining the engine (and/or transmission) oil, or the coolant. Oil or coolant that is very hot may scald you. If you can touch the engine sump/transmission/radiator (as applicable) without discomfort, the engine has probably cooled sufficiently to avoid scalding.

ALWAYS allow the engine to cool before working on it. Many parts of the engine become very hot in normal operation, and you may burn yourself badly.

ALWAYS keep brake fluid, antifreeze and other similar liquids away from your car's paintwork, as it may damage the finish (or may remove the paint altogether!).

ALWAYS keep toxic fluids, such as fuel, brake and transmission fluids and antifreeze off your skin, and never syphon them by mouth.

ALWAYS wear a mask when doing any dusty work, or where spraying is involved, especially when doing body repair and brake jobs.

ALWAYS clean up oil and grease spills – someone may slip and be injured.

ALWAYS use the right tool for the job. Spanners and screwdrivers which don't fit properly are likely to slip and cause injury.

ALWAYS get help to lift and handle heavy parts – it's never worth risking an injury.

ALWAYS take time over the jobs you take on. Plan out what you must do, and make sure that you have the right tools and spare parts. Follow the recommended steps, and check over the job once you're finished. Leave 'short-cuts' to the experts.

ALWAYS wear eye protection when using power tools such as drills, sanders and grinders, when using a hammer, and when working beneath your car.

ALWAYS use a barrier cream on your hands when doing dirty jobs, especially when in contact with fuel, oils, greases, and brake and transmission fluids. It will help protect your skin against infection, and will make the dirt easier to remove later. Make sure that your hands aren't slippery. The use of a suitable specialist hand cleaner will make dirt and grease easier to remove without causing infection or damage to skin. Long term or regular contact with used oils and fuel can be a health hazard.

ALWAYS keep loose clothing and long hair out of the way of any moving parts.

ALWAYS take off rings, watches, bracelets

VAUXHALL ASTRA & BELMONT

SERVICING

and neck chains before starting work on your car, especially the electrical system.

ALWAYS make sure that any jacking equipment or lifting tackle you use has a safe working load which will cope with what you intend to do, and use the equipment exactly as recommended by the manufacturers.

ALWAYS keep your work area tidy; someone may trip or slip on articles left lying around and be injured.

ALWAYS get someone to check up on you from time to time if you're working on the car alone, to make sure that you're alright.

ALWAYS do the work in a logical order, check that you've put things back together properly, and that everything is tightened as it should be.

ALWAYS keep children and animals away from an unattended car, and from the area where you're working.

ALWAYS park cars with catalytic converters away from materials which may burn, such as dry grass, oily rags, etc, if the engine has recently been run. Catalytic converters reach extremely high temperatures, and any such materials close by may catch fire.

REMEMBER that your safety, and that of your car and other people, rests with you. If you are in any doubt about anything, get specialist advice and help right away.

IF in spite of these precautions you injure yourself – get medical help immediately.

Asbestos

Some parts of your car, such as brake pads and linings, clutch linings and gaskets contain asbestos, and you should take appropriate precautions to avoid inhaling dust (which is hazardous to health) when working with them. If in doubt, assume that they **do** contain asbestos.

Fire

Remember at all times that petrol is highly flammable. Never smoke, or have any kind of naked flame around, when working on the car. But the risk doesn't end there – a spark caused by an electrical short-circuit, by two metal surfaces contacting each other, by careless use of tools, or even by static electricity built up in your body under certain conditions, can ignite petrol vapour, which in a confined space is highly explosive.

If a fuel leak is suspected, try to find the cause; seek advice as soon as possible, and never risk fuel leaking onto a hot engine or exhaust. Catalytic converters (where fitted) run at extremely high temperatures, and therefore can be an additional fire hazard – observe the precautions outlined previously in this Section.

It's recommended that a fire extinguisher of a type suitable for fuel and electrical fires is kept handy in the garage or workplace at all times.

Never try to extinguish a fuel or electrical fire with water.

Fumes

Certain fumes are highly toxic, and can quickly cause unconsciousness and even death if inhaled to any extent, especially if inhalation takes place through a lighted cigarette or pipe. Petrol vapour comes into this category, as do the vapours from certain solvents such as trichloroethylene. Any draining or pouring of such fluids should be done in a well-ventilated area.

When using cleaning fluids and solvents, read the instructions carefully. Never use materials from unmarked containers – they may give off poisonous vapours.

Never run a car engine in an enclosed space such as a garage. Exhaust fumes contain carbon monoxide which is extremely poisonous; if you need to run the engine, always do so in the open air, or at least have the rear of the car outside the workplace. Although cars fitted with catalytic converters produce far less toxic exhaust gases, the above precautions should still be observed.

If you're fortunate enough to have the use of an inspection pit, never run the engine while the car is standing over it; the fumes, being heavier than air, will concentrate in the pit with possibly lethal results.

The battery

Never short across the two poles of the battery! (A battery is shorted by connecting the two terminals directly to each other, and this can happen accidentally when working under the bonnet with metal tools.) The heavy discharge caused will create 'gassing' of hydrogen from the battery, and this is highly explosive. Shorting across a battery may also cause sparks, and the combination can cause the battery to explode, with potentially lethal results. A conventional battery will normally be giving off a certain amount of hydrogen all the time, so for the same reasons given above, never cause a spark or allow a naked flame close to it.

Batteries which are sealed for life require special precautions which are normally outlined on a label attached to the battery. Such precautions usually relate to battery charging and jump starting from another vehicle (for details of jump starting refer to 'Breakdowns' on page 52).

If possible, loosen the filler plugs or covers when charging the battery from an external source. Don't charge at an excessive rate, or the battery may burst. Special care should be taken with the use of high charge-rate boost chargers to prevent the battery from overheating.

Take care when topping-up and when carrying the battery. The battery contains dilute sulphuric acid which is very corrosive. It will burn and may cause long-term damage if in contact with skin, eyes or clothes. Similarly, corrosive deposits around the battery terminals may be harmful.

Always wear eye protection when cleaning the battery, to prevent the corrosive deposits from entering your eyes.

Mains electricity and electrical equipment

When using any electrical equipment which works from the mains, such as an electric drill or inspection light, always ensure that the appliance is correctly connected to its plug and that, where necessary, it's properly earthed. Always use an RCD (Residual Current Device, a safety device incorporating a circuit breaker) when using mains electrical equipment. Don't use such appliances in damp conditions, and take care not to create a spark or apply excessive heat close to fuel or fuel vapour. Also make sure that the appliances meet the relevant national safety standards.

Ignition HT voltage

A severe electric shock can result from touching certain parts of the ignition system, such as the spark plug HT leads, when the engine is running or being cranked, particularly if components are damp or the insulation is faulty. Where an electronic ignition system is fitted, the HT voltage is much higher than that used in a contact breaker system, and could prove fatal, especially to wearers of cardiac pacemakers.

Disposing of used engine oil

Used engine oil is a hazard to health and the environment, and should be disposed of safely and cleanly. Most local authorities provide a disposal site which will have a special tank for waste oil. If in doubt, contact your local council for advice on where you can dispose of oil safely – there may be a local garage who will allow you to use their specialised oil disposal tank free of charge. Remember that it's not just the oil which causes a problem, but empty containers, old oil filters, and oily rags, so these should be taken to your local disposal site too.

Do not under any circumstances pour engine oil down a drain, or bury it in the ground.

Jacking and vehicle support

The jack provided with the car is designed for emergency wheel changing, and should not be used for servicing and overhaul work. Instead, a more substantial workshop jack (trolley jack or similar) should be used. Whichever type is used, it's essential that additional safety support is provided by means of axle stands designed for this purpose. Never use makeshift means such as narrow wooden blocks or piles of house bricks, as these can easily topple or, in the case of bricks, disintegrate under the weight of the car. Further information on the correct positioning of the jack and axle stands is provided at the end of this Section.

If you don't need to remove the wheels, the use of drive-on ramps is recommended. Ensure that the ramps are correctly aligned with the wheels, and that the car is not driven too far along them, so that it promptly falls off the other end, or tips the ramps.

JACK AND AXLE STAND POSITIONS

When using a trolley jack or any other type of workshop jack to raise the car, the jack head should be placed under the special platforms located between the car jacking points and the wheel arch. To avoid damaging the underbody, it is recommended that a rubber pad or wooden block is positioned between the jack head and the body.

SERVICING 101

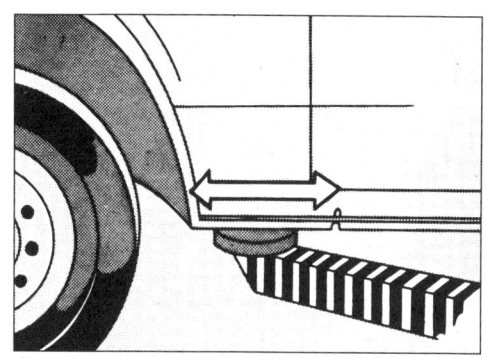

▲ *Jacking area (within arrows) for trolley jack*

Do not jack the car under the engine sump or any of the steering or suspension components.

Axle stands can be positioned under the same platforms. If necessary, place the trolley jack slightly to one side of the platform when jacking up the car, in order to leave room for the axle stand head when the car is lowered. This axle stand position is to be preferred to placing it under the car jacking points reserved for the car jack, as the sill contour is specially made to accept the car jack head, and might be damaged if a trolley jack is used.

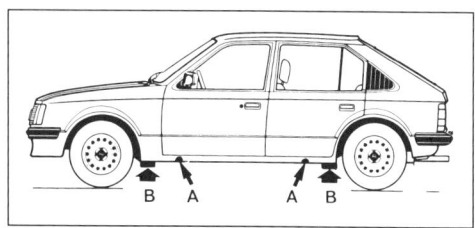

▲ *Jacking and support points*
  A *Jacking points for car jack only*
  B *Jacking/support platforms for trolley jack/axle stands*

BUYING SPARE PARTS

When buying spare parts such as oil or air filters, make sure that the correct replacement parts are obtained for your particular car. Many changes are made during the production run of any car, and when ordering spare parts, it will usually be necessary to know the year the car was built, the model type (eg 'GL or 'SRi') and engine size, and in some cases the Vehicle Identification Number (VIN). The VIN is stamped on a metal plate attached to the cross panel at the front of the engine compartment, and is also stamped in the floor beneath the carpet next to the driver's seat. It also appears on the Vehicle Registration Document.

All of the components required for servicing will be available from an official Vauxhall dealer, but most parts should also be available from good motor accessory shops and motor factors.

Identifying engines

There are two types of engine fitted to the Astra/Belmont – the overhead valve (OHV) type 1.2 litre engine, and the overhead camshaft (OHC) type 1.3, 1.4, 1.6, 1.8 & 2.0 litre engines. On the overhead valve engine, the distributor is located vertically on the front of the engine block with its cap facing upwards, while on the overhead camshaft engine, the distributor is located horizontally on the end of the camshaft, with its cap facing the battery. OHC engine models are usually stamped 'OHC' on top of the engine, on the camshaft cover. The 2.0 litre 16-valve engine is clearly marked 'VAUXHALL DOHC 16V' on top of the engine. To distinguish between a fuel injection engine and a carburettor engine, look for the air cleaner position – on a carburettor engine it is located centrally over the engine and is pancake-shaped, whereas on fuel injection engines it is located on the right-hand side of the engine compartment, and is box-shaped.

102 SERVICING

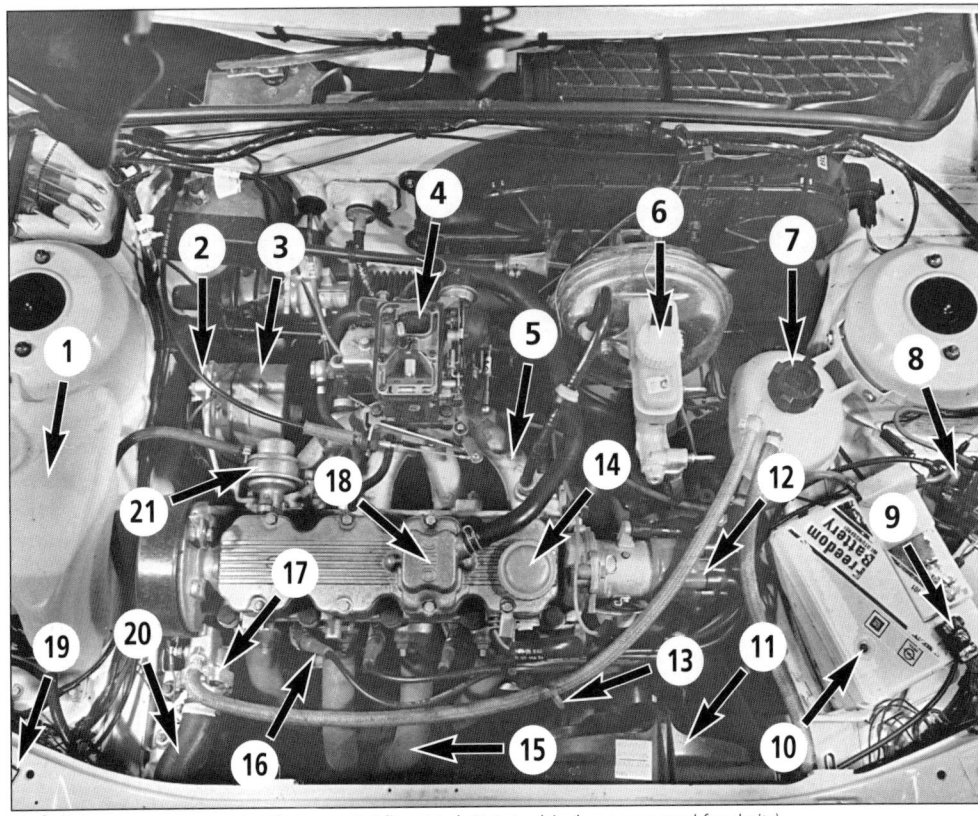

▲ *Underbonnet component locations on a 1.6 litre 'Mark 1' Astra (air cleaner removed for clarity)*

1. Windscreen and headlamp washer reservoir
2. Accelerator cable
3. Alternator
4. Carburettor
5. Inlet manifold
6. Brake fluid reservoir
7. Coolant expansion tank with filler/pressure cap
8. Ignition coil
9. Battery earth terminal
10. Battery condition indicator
11. Radiator cooling fan shroud
12. Distributor cap and HT leads
13. Engine oil level dipstick
14. Engine oil filler cap
15. Exhaust manifold
16. Number 1 spark plug HT lead
17. Thermostat housing
18. OHC engine identification
19. Vehicle Identification Number (VIN) plate
20. Radiator top hose
21. Fuel pump

VAUXHALL ASTRA & BELMONT

SERVICING 103

▲ *Underbonnet component locations on a 1.6 litre 'Mark 2' Astra – air cleaner removed for clarity (Belmont similar)*

1. Brake fluid reservoir
2. Alternator
3. Carburettor
4. Inlet manifold
5. Windscreen washer reservoir
6. Manual gearbox oil filler plug
7. Coolant expansion tank with filler/pressure cap
8. Ignition coil
9. Battery earth terminal
10. Battery condition indicator
11. Radiator bottom hose
12. Radiator cooling fan shroud
13. Distributor cap and HT leads (waterproof cover fitted)
14. Engine oil level dipstick
15. Engine oil filler cap
16. Exhaust manifold
17. Vehicle Identification Number (VIN) plate
18. Number 1 spark plug HT lead
19. OHC engine identification
20. Radiator top hose
21. Thermostat housing
22. Accelerator cable
23. Choke cable

VAUXHALL ASTRA & BELMONT

SERVICING

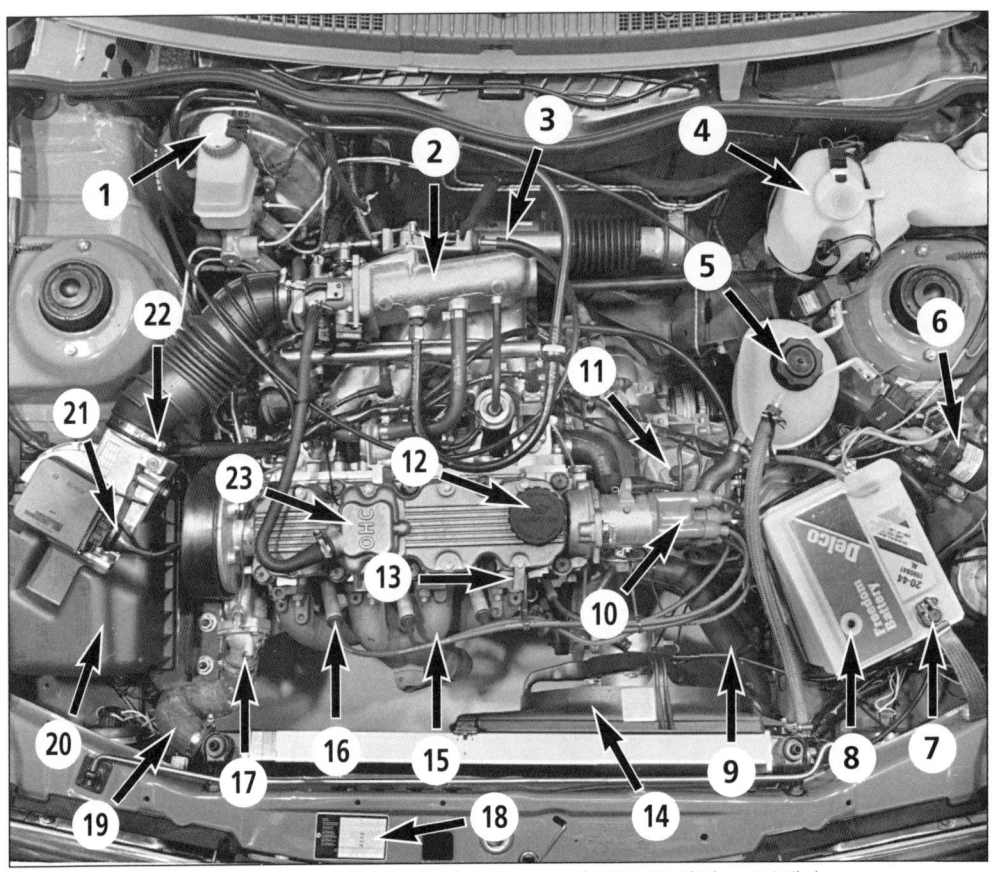

▲ *Underbonnet component locations on a 1.8 litre 'Mark 2' Astra with fuel injection (Belmont similar)*

1. Brake fluid reservoir
2. Throttle housing and inlet manifold
3. Accelerator cable
4. Windscreen washer reservoir
5. Coolant expansion tank with filler/pressure cap
6. Ignition coil
7. Battery earth terminal
8. Battery condition indicator
9. Radiator bottom hose
10. Distributor cap and HT leads
11. Manual gearbox oil filler plug
12. Engine oil filler cap
13. Engine oil level dipstick
14. Radiator cooilng fan shroud
15. Exhaust manifold
16. Number 1 spark plug HT lead
17. Thermostat housing
18. Vehicle Identification Number (VIN) plate
19. Radiator top hose
20. Air cleaner
21. Airflow sensor plug
22. Air trunking securing clip
23. OHC engine identification

VAUXHALL ASTRA & BELMONT

SERVICING 105

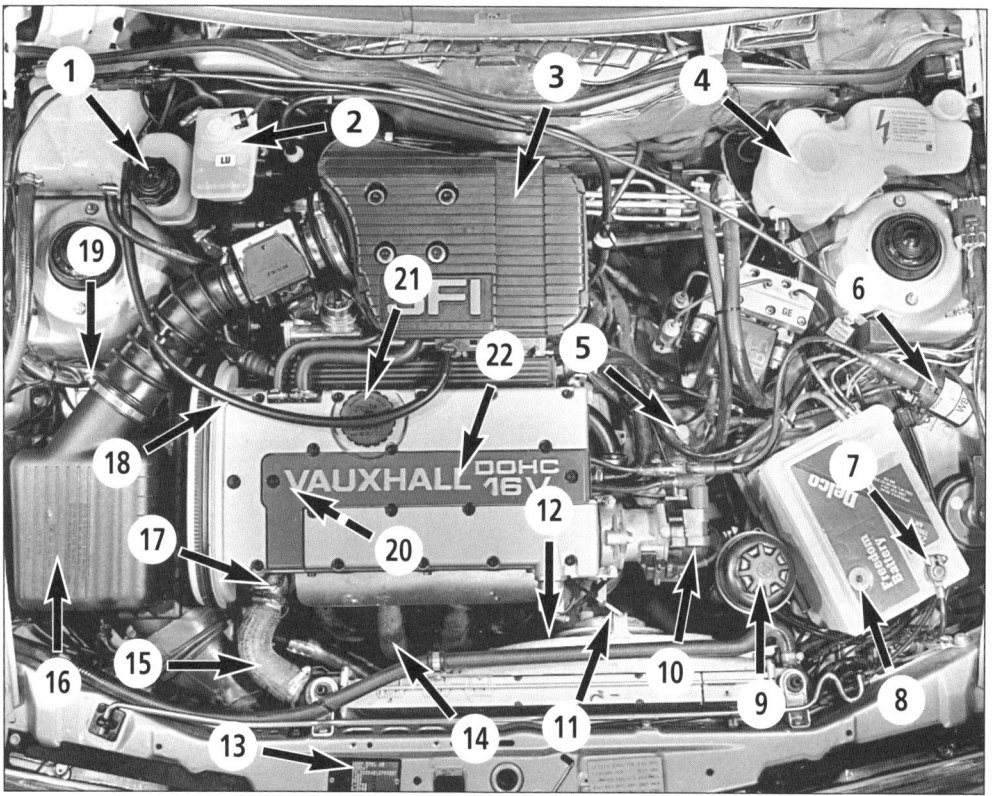

▲ *Underbonnet component locations on a 'Mark 2' 2.0 litre 16-valve Astra GTE*

1. Coolant expansion tank with filler/pressure cap
2. Brake fluid reservoir
3. Throttle housing (inlet manifold below)
4. Windscreen washer fluid reservoir
5. Manual gearbox oil filler plug
6. Ignition coil
7. Battery earth terminal
8. Battery condition indicator
9. Power steering fluid reservoir
10. Distributor cap
11. Engine oil level dipstick
12. Radiator cooling fan shroud
13. Vehicle Identification Number (VIN) plate
14. Exhaust manifold
15. Radiator top hose
16. Air cleaner
17. Thermostat housing
18. Accelerator cable
19. Air trunking securing clip
20. Number 1 spark plug (below HT lead cover plate)
21. Engine oil filler cap
22. Spark plug HT lead cover plate

SERVICING

SERVICE TASKS

The following pages provide easy-to-follow instructions to enable you to carry out the regular service tasks recommended by the manufacturer.

Every 250 miles (400 km) or weekly

Refer to 'Regular checks' and carry out all the checks described.

Every 6000 miles (10 000 km) / 6 months – whichever comes first, OR every 9000 miles (15 000 km) / 1 year – whichever comes first

Note: *The shorter time/mileage interval shown above applies to models produced up to the end of September 1982. Models from October 1982 on use the longer interval for all these service tasks. However, on models produced from October 1982 up to the end of September 1986, the engine oil and filter should be renewed every 9000 miles (15 000 km) or 6 months – whichever comes first.*

ENGINE COMPARTMENT

Check all hoses and pipes, and the surfaces of all components for signs of fluid leakage, corrosion or deterioration

If there's any sign of leaking fluid, try to find the source of the leak, and seek advice if necessary. It can be difficult to identify leaking fluids, but if there's obviously a major leak, or if you suspect even a slight brake fluid or petrol leak, don't drive the car until the problem has been investigated by someone suitably qualified.

Check all rubber or fabric hoses for signs of cracking, damage due to rubbing on other components, and general deterioration. It's sensible to have any suspect hoses renewed as a precaution against possible failure.

Check for obvious signs of corrosion on metal pipes, hose clips, and component joints, which may indicate a fluid leak. Any badly-corroded components should be cleaned and, if necessary, renewed. Pay particular attention to metal fluid pipes, and have them renewed if they're badly pitted or corroded.

ENGINE

Renew the engine oil and filter

Note: *The following items will be required for this task:*
- *A suitable quantity of engine oil of the correct type – refer to* **'Service specifications'** *on page 73*
- *New oil filter (of the correct type for your particular car)*
- *New oil drain plug sealing washer (depending on the condition of the old one)*
- *Suitable container to catch the old oil as it drains (make sure that the capacity of the container is larger than the oil capacity of the engine – refer to* **'Service specifications'** *on page 74).*
- *Oil filter wrench – for removal of old oil filter*
- *Small quantity of clean rag*
- *Suitable spanner or socket to fit oil drain plug*

The oil should be drained when the engine is warm, immediately after a run. Ideally, the front of the car should be jacked up to improve access. If the front of the car is to be raised, apply the handbrake, then jack up the front of the car and support it securely on axle stands (refer to 'Jacking and vehicle support' in 'Safety first!' on page 100 for details of where to position the jack and axle stands).

Position a suitable container under the engine sump to catch all the oil which will be drained.

Remove the oil filler cap from the top of the engine, then working under the car, unscrew the drain plug from the engine sump using a suitable spanner or a socket. The drain plug is located on the bottom rear edge of the sump. Some oil will be released before the drain plug is removed completely, so take precautions against scalding, as the oil will be hot. Try not to drop the drain plug when removing it.

Allow the oil to drain for at least 15 minutes.

SERVICING 107

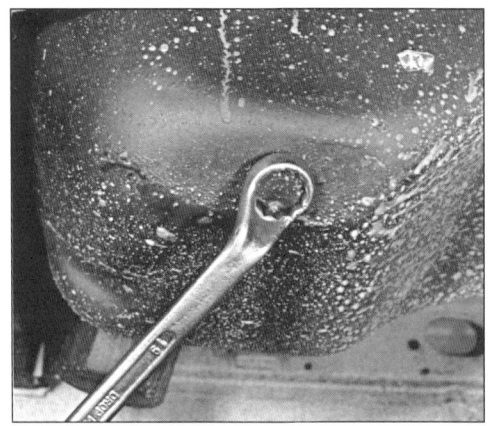

▲ *Unscrewing the engine oil drain plug*

Check the condition of the oil drain plug sealing washer, and renew it if necessary.

When the oil has finished draining, clean the drain plug, washer, and the mating face of the sump, then refit and tighten the drain plug. Tighten the drain plug securely, but there's no need to strain yourself – remember that it could be you who has to undo it, next time!

Position a suitable container under the oil filter. On the 1.2 litre OHV engine, the filter is located on the front of the engine, below the distributor. On the 1.3 & 1.4 litre engines, the oil filter is located on the front of the engine next to the exhaust downpipes – be careful not to burn yourself on the hot exhaust when removing the filter. On all other engines, the filter is located on the rear of the engine, below the alternator lower mounting bracket.

Unscrew the filter, using a suitable chain or strap wrench if necessary to loosen it. Make sure that the oil filter rubber sealing ring is removed with the old filter, and that it is not left on the engine.

Wipe clean the oil filter mounting flange on the engine. Smear a little clean engine oil on the sealing ring of the new oil filter, then screw the filter onto the engine, and tighten it *by hand only*. If no instructions are provided with the filter, tighten it until the sealing ring touches the mounting face on the engine, then tighten it a further three-quarters of a turn.

Where applicable, lower the car to the ground and, with the car parked on level ground, fill the engine with the correct quantity and grade of oil through the filler on top of the engine. Fill the engine slowly, over a period of several minutes, until the level reaches the **'MAX'** or upper mark on the dipstick.

Make sure that the oil filler cap is fitted on completion.

Don't race the engine when starting it for the first time after an oil change, as the oil will take a few seconds to circulate. There may be a delay before the oil pressure warning light goes out, as the engine lubrication system fills with oil.

Run the engine, and check for leaks from the filter and the drain plug, then stop the engine and check the oil level. Refer to *'Regular checks'* for details of how to check the oil level. Since some oil will have filled the new oil filter, it will be necessary to top-up the level a little.

Dispose of the old engine oil safely (refer to *'Safety first!'* on page 100). **Don't** pour it down a drain.

COOLING SYSTEM

Check the coolant antifreeze concentration

Using an antifreeze hydrometer, withdraw some coolant and read off the concentration on the hydrometer float.

Refer to *'Regular checks'* on page 77 for the correct concentration, and if necessary adjust by adding antifreeze. If this will raise the coolant level above the maximum mark, first withdraw some of the coolant using the hydrometer. If a large amount must be

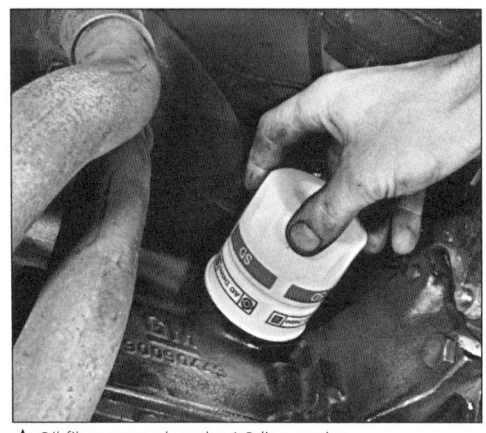

▲ *Oil filter removal on the 1.3 litre engine*

108 SERVICING

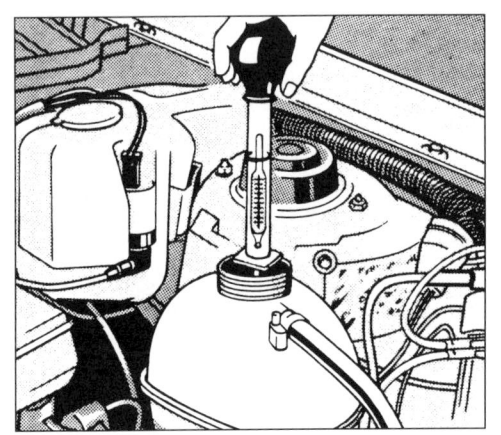

▲ *Checking the antifreeze concentration with a hydrometer*

removed, drain off some coolant, as described in the '*Every 2 years, regardless of mileage*' service Section on page 123.

FUEL AND EXHAUST SYSTEMS

Clean the fuel pump filter screen (early carburettor models)

On 1.2 litre OHV engines, the fuel pump is located on the front facing side of the cylinder block, at the right-hand end (when viewed from the driver's seat). On OHC engines, it is located on the rear-facing side of the engine, in front and to the right of the carburettor. First unscrew and remove the screw from the centre of the fuel pump cover.

Withdraw the cover, then turn it over and remove the rubber sealing ring and filter screen.

Wipe out the cover and pump recess, and wash the filter screen in clean fuel if it is clogged or partially choked with dirt.

Refit the filter screen and rubber seal to the cover, then refit the cover to the pump body. Tighten the screw, but take care not to over-tighten it, as it is quite easy to strip the threads in the pump body.

Lubricate carburettor throttle and choke linkages

Have an assistant work the accelerator pedal, and then operate the manual choke control, while you look in the engine compartment to identify the cables and linkages. Apply a little engine oil to the throttle linkage, and where applicable the choke linkage, on the carburettor. Operate both the accelerator and the choke immediately afterwards to allow the oil to penetrate the linkages.

IGNITION SYSTEM

Clean or renew the contact breaker points (pre-October 1982 1.3 litre models, and all 1.2 litre models)

Note: *A set of feeler gauges will be required for this task.*

To get at the contact breaker points, you'll need to remove the distributor cap. Start by prising apart the waterproof cover, if fitted,

▲ *Removing the fuel pump cover, filter screen and rubber sealing ring*

▲ *Removing the cap cover from the distributor (1.3 litre engine, Bosch distributor)*

VAUXHALL ASTRA & BELMONT

SERVICING 109

▲ *Removing the distributor cap screw (1.2 litre engine, Delco-Remy distributor)*

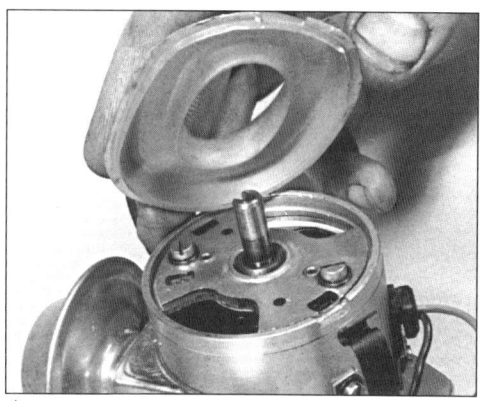

▲ *Removing the plastic cover (1.3 litre engine, Bosch distributor)*

▲ *Removing the rotor arm (1.3 litre engine, Bosch distributor)*

▲ *Removing the rotor arm (1.2 litre engine, Delco-Remy distributor)*

and removing it. This exposes the cap itself, which can then be removed by releasing the two spring clips or, on Delco-Remy distributors, by removing the two screws.

Pull off the rotor arm and remove the plastic cover below it, where fitted.

On 1.3 litre engines, remove the two screws which secure the distributor upper bearing plate, then remove the plate and the lower retaining ring.

The contact points can now be examined for condition. Gently prise the contacts apart and examine the condition of their faces. If they are rough, pitted or dirty, it will be necessary to remove them for resurfacing, or for new points to be fitted.

On the Bosch distributor, remove the LT

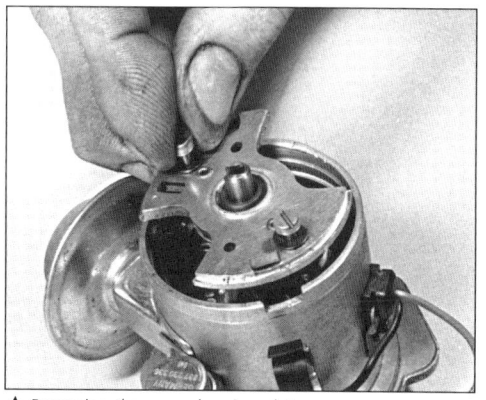

▲ *Removing the upper bearing plate (1.3 litre engine, Bosch distributor)*

VAUXHALL ASTRA & BELMONT

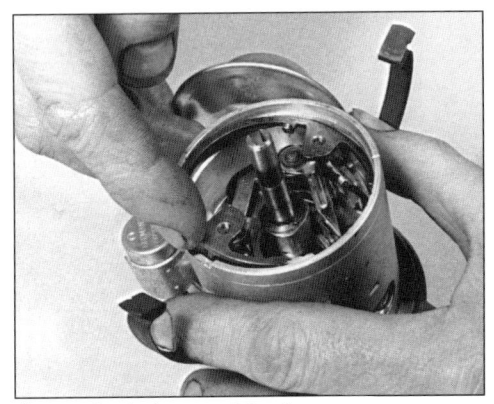

SERVICING

spade connector, then remove the setscrew retaining the contact breaker mechanism to the baseplate, and lift out the complete assembly.

On the Delco-Remy distributor, unscrew and remove the setscrew retaining the contact mechanism to the baseplate. Ease the moving contact arm spring from the insulator assembly on the fixed contact plate, and remove the condenser and LT spade terminals. Lift off the moving contact arm and spring, followed by the fixed contact plate.

It is possible to reface the contact points where required, using a fine carborundum stone or emery paper, and taking care to keep the points faces flat and parallel to each other. However, if the points show signs of excessive burning or pitting, it is strongly recommended

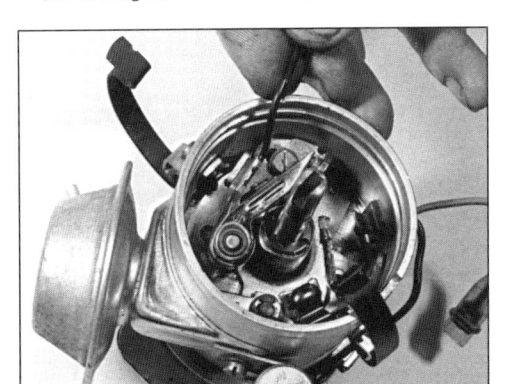

▲ Removing the lower retaining ring
(1.3 litre engine, Bosch distributor)

▲ Disconnecting the LT spade connector
(1.3 litre engine, Bosch distributor)

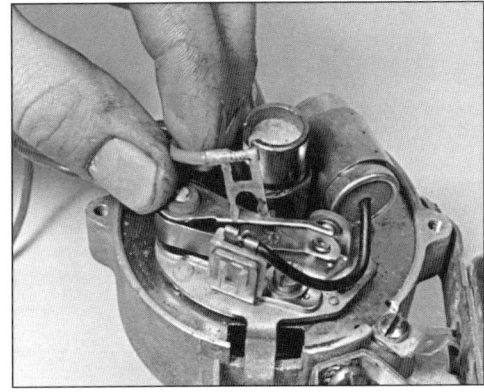

▲ Disconnecting the LT and condenser leads
(1.2 litre engine, Delco-Remy distributor)

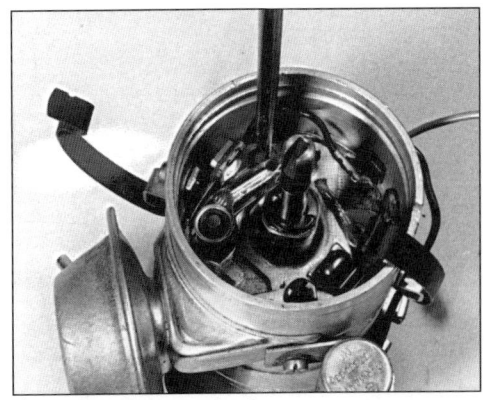

▲ Removing the contact breaker screw
(1.3 litre engine, Bosch distributor)

▲ Removing the contact breaker screw (1.2 litre engine, Delco-Remy distributor)

VAUXHALL ASTRA & BELMONT

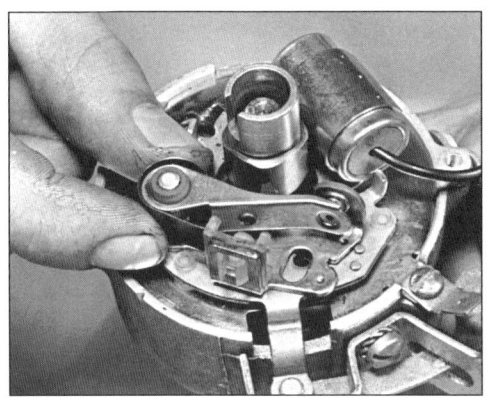

SERVICING 111

▲ *Removing the contact breaker set (1.2 litre engine, Delco-Remy distributor)*

that they are replaced with a new set.

Clean the faces of new points with methylated spirit before fitting.

Refit the contact points using a reversal of the removal procedure. When fitting the Delco-Remy type points on the 1.2 litre engine, ensure that the LT and condenser leads are fitted in the correct order on the insulator post.

▲ *LT and condenser leads fitted in the correct order (1.2 litre engine, Delco-Remy distributor)*

On the 1.3 litre engine, it is essential that the bearing plate is refitted before adjusting the points gap, otherwise the setting will be incorrect – apply a little engine oil to the bearing in the upper plate before refitting it. Take care not to contaminate the point faces with oil when fitting them.

It is now necessary to adjust the points. Unfortunately, on the 1.3 litre engine, because the top plate has to be in position, you'll have to work through the holes in the plate to make the adjustment.

Turn the engine by pushing the car along in top gear (manual gearbox), or by using a spanner on the crankshaft pulley bolt (the crankshaft pulley bolt is located on the lower right-hand end of the engine – viewed from the driver's seat – in the centre of the crankshaft pulley). Turning the engine will be much easier if the spark plugs are temporarily removed (refer to page 112), but be sure to mark the spark plug HT leads for position before removing them. Turn the engine until the heel of the moving contact is resting on the peak of one of the cam lobes (the cam lobes are the four raised 'corners' on the distributor shaft). In this position, the gap between the contact point faces is at its maximum, and it is in this position that adjustment is made.

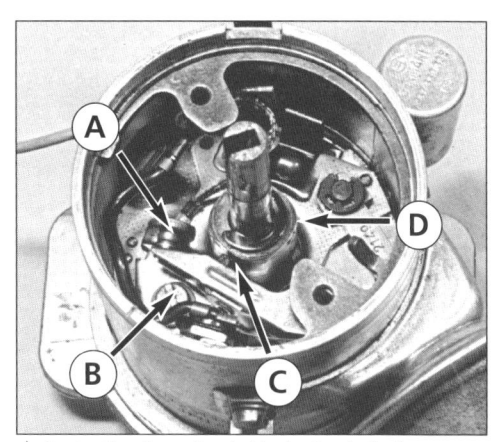

▲ *Contact breaker points inside the Bosch distributor (Delco-Remy similar)*
  A Contact breaker points
  B Securing screw
  C Heel of moving contact
  D Distributor shaft cam lobe

Insert a clean feeler blade of the correct thickness between the contact point faces (refer to *'Service specifications'* on page 75 for the correct gap). Slacken the points securing screw, then gently lever the stationary half of the contact set to increase or reduce the gap,

VAUXHALL ASTRA & BELMONT

SERVICING

▲ *Adjusting the points gap
(1.2 litre engine, Delco-Remy distributor)*

▲ *Oiling the distributor felt pad
(1.2 litre engine, Delco-Remy distributor)*

that the carbon contact in the centre of the cap is not excessively worn, make sure that it's free to move, and that it protrudes from its holder – if not, renew the cap.

Snap together the waterproof cover (when fitted), and start the engine just to make sure that the new points are correctly fitted. Contamination of their faces, perhaps with oil from a dirty feeler blade, could prevent the engine running.

On modern engines, setting the points gap as just described must be regarded as a preliminary adjustment. For best results, the dwell angle should now be measured. Since altering the points gap will also affect the ignition timing, this too should be checked after fitting a new set of points. For both these tasks, refer to the Owners Workshop Manual.

Check the spark plugs and renew if necessary

Note: *A suitable spark plug spanner and a spark plug gap adjustment tool or a set of feeler gauges will be required for this task.*

It is preferable to remove the spark plugs when the engine is cold.

On the 2.0 litre 16-valve engine, remove the two Allen screws which secure the spark plug lead cover, then remove the cover.

If necessary, mark the HT leads using tape or sticky labels, to make sure that they're refitted in their correct positions – on the 16-valve engine this is unnecessary, as all of the leads

until the feeler blade is a firm sliding fit, ie neither tight nor slack. Tighten the securing screw and recheck the gap.

On 1.2 litre engines, apply one or two drops of engine oil to the felt pad in the top of the distributor shaft.

Refit the plastic cover and the rotor arm. Before refitting the distributor cap, clean it thoroughly both inside and outside. Examine the inside of the distributor cap, and if any of the four HT lead segments appear badly burnt or pitted, or if there are any signs of hairline cracks in the plastic, renew the cap. Make sure

▲ *Removing the spark plug lead cover on the 16-valve engine*

VAUXHALL ASTRA & BELMONT

SERVICING 113

are of different lengths, and it is impossible to mix them up.

If the plugs are removed one at a time, as described in the following procedure, there will be no chance of getting the leads wrong – work from one end of the engine to the other.

Starting with the first spark plug, look to see if a metal shroud is fitted at the end of the HT lead. If so, remove the lead and shroud together by pulling on the metal shroud – this may be rather stiff, but careful pulling and twisting should free it. Do not use excessive force, or the spark plug will be damaged, and a new one would then be needed.

If there is no metal shroud, remove the HT lead by pulling on the rubber insulator at the end of the lead, not on the lead itself.

Before removing the spark plug, brush or blow any grit from the area around the plug recess, to avoid it dropping into the engine as the plug is removed.

Using a spark plug spanner, unscrew the spark plug and remove it from the engine. The use of a spark plug spanner with a rubber insert will be helpful to remove the plug, especially on the 2.0 litre 16-valve engine.

If the spark plug electrodes are heavily fouled (very dirty), it may be advisable to renew all four plugs, especially if poor starting has been experienced. Spark plug cleaning should only

▲ *Disconnecting a spark plug lead with metal shroud (except on the 2.0 litre 16-valve engine)*

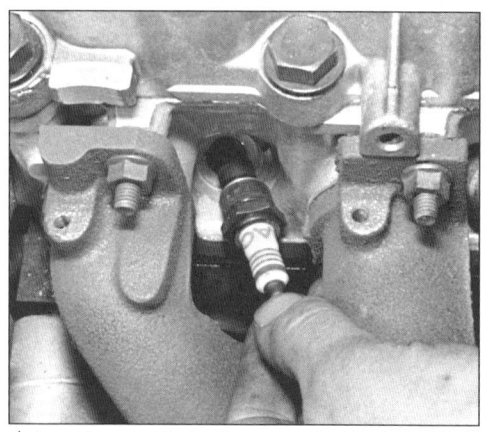

▲ *Removing a spark plug (except on the 16-valve engine)*

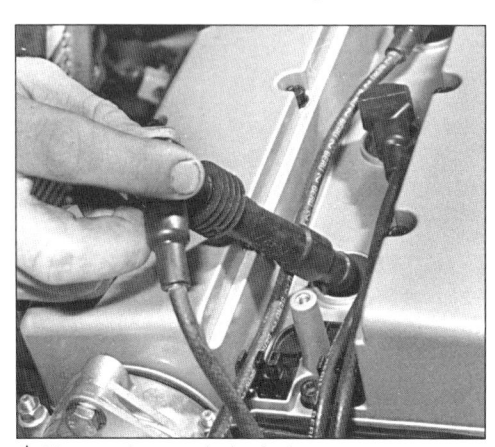

▲ *Disconnecting a spark plug lead (2.0 litre 16-valve engine)*

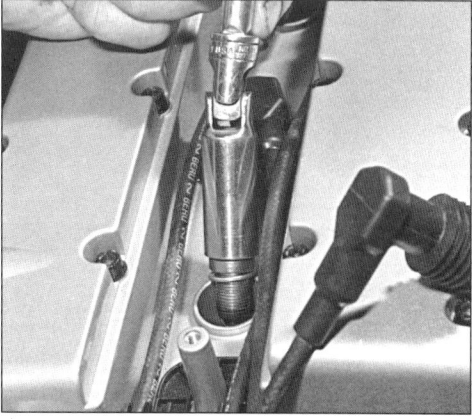

▲ *Removing a spark plug on the 2.0 litre 16-valve engine, using a spark plug spanner with a rubber insert*

VAUXHALL ASTRA & BELMONT

SERVICING

be carried out by someone suitably qualified using specialised equipment, otherwise the plugs may easily be damaged.

If the plug appears to be in good condition, check the electrode gap as follows before refitting it (refer to *'Service specifications'* on page 75 for the correct electrode gap). The size of the spark plug electrode gap is extremely important, and it can drastically affect the performance of the engine.

Measure the gap between the electrodes using a spark plug gap adjustment tool, or a feeler gauge.

▲ *Checking a spark plug electrode gap*

▲ *Adjusting a spark plug electrode gap*

If adjustment is required, carefully bend the **outer** electrode until the correct gap is obtained. **Never** try to bend the centre electrode.

Before fitting the plug, check that the 'nut' on the threaded terminal at the top of the plug is tight.

Make sure that the plug threads and the seating areas in the cylinder head are clean, and apply a little light oil to the plug threads. Screw the plug in by hand initially.

Using a spark plug spanner, tighten the plug as follows. With new plugs, screw in until the sealing washer contacts the cylinder head, then tighten no more than a quarter-turn. With plugs which have been used before, tighten by (approximately) a twelfth of a turn after the washer contacts the cylinder head, as the washer will already have been compressed. **Do not** overtighten the plugs.

Refit the HT lead (with metal shroud, where applicable), making sure that it is a secure fit over the end of the plug – there should be a soft 'click' as the lead is connected.

Repeat the above procedures for the remaining three spark plugs in turn, removing only one HT lead at a time, to avoid confusion.

On completion, check once more that all four HT leads are securely fitted. On the 2.0 litre 16-valve engine, refit the spark plug cover and tighten the screws after all the plugs and leads have been refitted.

MANUAL GEARBOX

Check and if necessary top-up the gearbox oil level

The level plug is located just to the rear of the left-hand driveshaft on the differential housing on 1.2 & 1.3 litre models, and just to the rear of the right-hand driveshaft on the differential housing on other models (as viewed facing forwards).

Access to the level plug is only possible from under the car, so it is necessary to raise the car and support on axle stands, or alternatively position the car over an inspection pit or ramp. Whichever method is used, it is important that the car is level.

Unscrew the level plug and check that the oil level is up to the bottom of the plug hole by inserting a finger or a suitable length of wire. On the 2.0 litre 16-valve model, the level

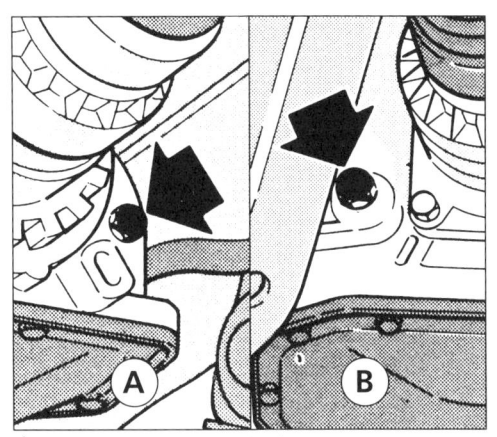

SERVICING 115

▲ Manual gearbox oil level plug location
A 1.2 & 1.3 litre models
B All other models

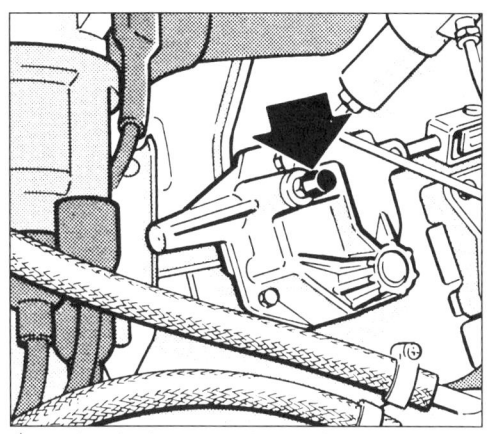

▲ Breather plug location on the top of the gearbox

should be 5.0 mm (0.2 in) below the bottom of the plug hole, and to check it, a dipstick should be made out of a piece of bent wire, with a mark 5.0 mm below the bend.

If topping-up is required, unscrew and remove the breather plug from the top of the gearbox and pour in the specified grade of oil (refer to 'Service specifications'). Do not under any circumstances top-up the gearbox through the level hole in the differential housing.

Note that if the gearbox requires repeated topping-up, there must be a leak somewhere,

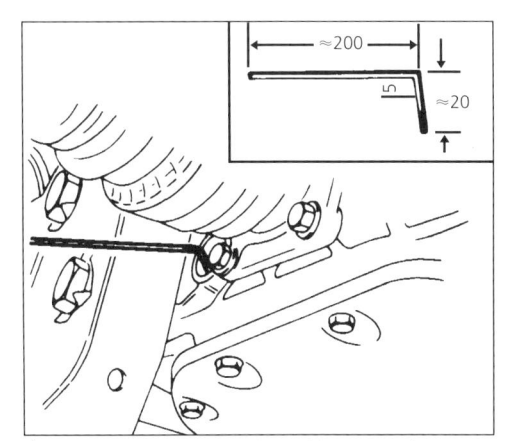

▲ Topping-up the manual gearbox oil level through the breather hole

which should be attended to as soon as possible. Refit and tighten the filler/breather and level plugs on completion.

AUTOMATIC TRANSMISSION

Check and if necessary top-up the automatic transmission fluid level

The automatic transmission fluid can be checked with the transmission hot or cold; 'hot' being defined as after motorway driving of at least 10 to 15 miles (15 to 20 km) or equivalent driving. Have the engine running, and the gear selector in position 'P' when making the check.

Withdraw the dipstick from its tube (located

▲ Checking the manual gearbox oil level on 2.0 litre 16-valve models
Inset shows dimensions of wire dipstick in mm

VAUXHALL ASTRA & BELMONT

SERVICING

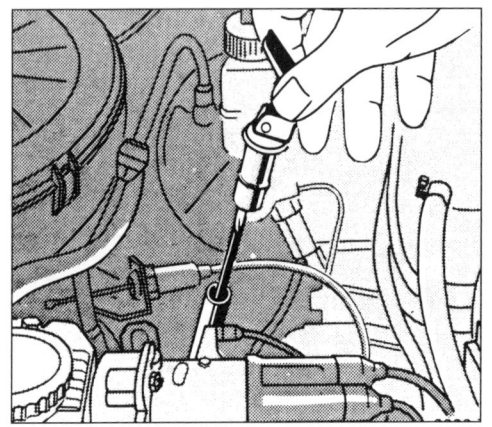

▲ *Automatic transmission dipstick location*

to the rear of the distributor), and wipe it dry with a clean lint-free rag.

Fully re-insert the dipstick, then withdraw it again and check the fluid level, which should be at or near the 'MAX' mark. Note that each side of the dipstick has a different marking – the 'cold' side is marked **+ 20°C** and the 'hot' side is marked **+ 94°C**. Make sure you refer to the correct side.

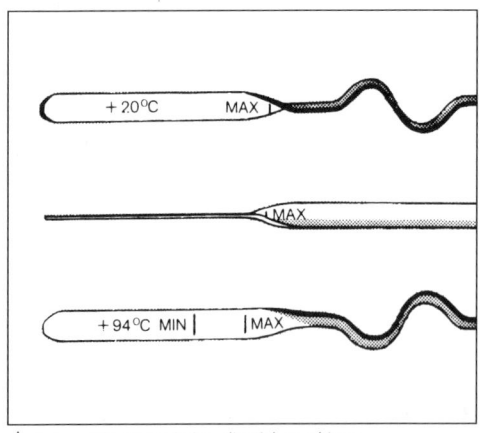

▲ *Automatic transmission dipstick markings*

If topping-up is necessary, use only the specified fluid type (refer to *'Service specifications'* on page 73). Top-up by pouring the fluid into the dipstick tube. Take care not to overfill – the level **must not** exceed the 'MAX' mark. Note that 5.0 mm (0.2 in) on the dipstick represents approximately 0.25 litre (0.44 pint).

If there's a need for regular topping-up, it's likely that there's a fluid leak, and the problem should be referred to a Vauxhall dealer.

SUSPENSION AND STEERING

Check the tightness of the roadwheel bolts

Using the wheelbrace from the car toolkit (refer to *'Changing a wheel'* on page 48), check the tightness of all roadwheel bolts. Don't over-tighten them, or you may not be able to undo them in the event of a puncture.

BODYWORK

Lubricate all hinges, door locks, and the bonnet release mechanism

Use general-purpose light oil for the door locks, and a suitable general-purpose grease for the hinges and bonnet release mechanism.

Don't use an excess of lubricant, or it may find its way onto other components, or onto the clothing of the car's occupants!

ELECTRICAL SYSTEM

Check the condition and adjustment of the alternator drivebelt, and (where fitted) the power steering pump drivebelt

Note: *On 1.2 litre engines, the drivebelt runs in the crankshaft, water pump and alternator pulleys. On all other engines, the belt only runs in the crankshaft and alternator pulleys, as the water pump is driven by the toothed camshaft belt. However, the removal, refitting and adjustment procedures on all models are similar.*

Examine the drivebelt(s) for signs of cracking, obvious wear, or contamination, and renew if necessary.

To check the tension of the drivebelt, press on it midway between the alternator/power steering pump and the crankshaft pulley. The deflection of the drivebelt should be approximately 13.0 mm (0.5 in) under moderate finger or thumb pressure.

To adjust the alternator drivebelt tension, slacken the alternator mounting and adjuster

VAUXHALL ASTRA & BELMONT

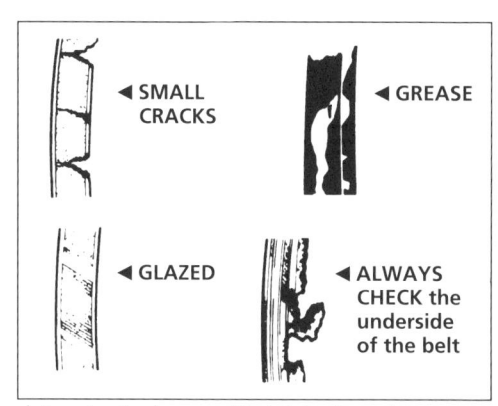

SERVICING 117

Examine the alternator drivebelt for signs of wear, deterioration or contamination

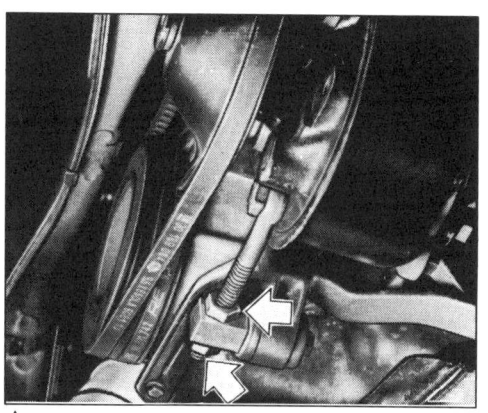

- ◄ SMALL CRACKS
- ◄ GREASE
- ◄ GLAZED
- ◄ ALWAYS CHECK the underside of the belt

Power steering pump adjuster locknuts – arrowed

link bolts/nuts just enough to enable the alternator to be moved and be held in the new position. Move the alternator towards or away from the engine as necessary to give the correct tension. Once the tension is correct, tighten the alternator mounting and adjuster link bolts/nuts.

Adjusting the alternator drivebelt tension (OHC engine shown)

To adjust the power steering pump drivebelt tension, slacken the mounting and adjuster bolts, then loosen the adjuster locknuts and turn them as necessary to reposition the pump and give the correct tension. Tighten the nuts and bolts on completion.

To renew a drivebelt, slacken all the mounting and adjuster bolts/nuts, move the alternator/power steering pump towards the engine, then remove the old drivebelt(s) from the pulleys. Note that on models with power steering, the power steering pump drivebelt must be removed first before the alternator drivebelt can be removed.

Fit the new drivebelt(s), and tension as already described. Run the engine for approximately ten minutes, then recheck the tension.

CAR UNDERSIDE

Check all hoses and pipes, and the surfaces of all components for signs of fluid leakage, corrosion, damage or deterioration

Refer to the *'ENGINE COMPARTMENT'* checks, at the beginning of the *'6000/9000 miles (10 000/15 000 km)'* service Section on page 106.

In addition, check the exhaust system and the fuel tank for any sign of damage or serious corrosion (light surface rust is to be expected).

Check the exhaust mountings to make sure that they're secure.

Check the visible suspension and steering components for obvious signs of damage or wear. In particular, check the steering gear rubber bellows and the tie-rod and suspension ball joint rubber dust covers for any damage or cracks.

VAUXHALL ASTRA & BELMONT

SERVICING

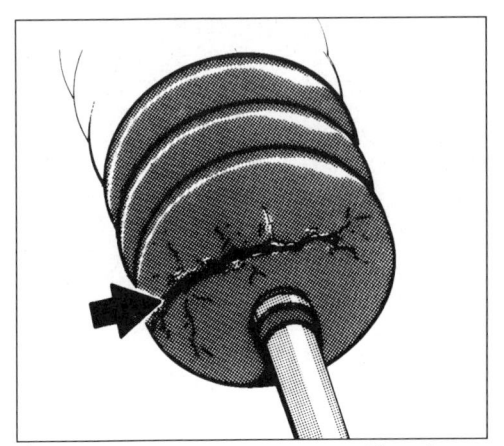

▲ *Check the driveshaft and steering gear rubber gaiters for splits (arrowed)*

Check the driveshafts for obvious signs of distortion or damage.

Pay particular attention to the driveshaft and steering gear rubber gaiters, which should be renewed if they're split, or if there are any obvious signs of lubricant leakage.

Check the brake hydraulic hoses and pipes for damage or corrosion.

ADDITIONAL TASKS

Note that as well as the tasks described in the preceding paragraphs, the additional tasks given in the *'Service schedule'* chart on pages 90 to 97 should also be carried out at this service interval. These tasks require more detailed explanation, or the use of special tools, and are considered beyond the scope of this Handbook. For details of these additional tasks, refer to the Owners Workshop Manual.

ROAD TEST

Check the operation of all instruments and electrical equipment

Make sure that all instruments read correctly, and switch on all electrical equipment in turn to check that it functions properly.

Check for any abnormalities in the steering, suspension, handling or road 'feel'

Drive the car and check that there are no unusual vibrations or noises.

Check that the steering feels positive, with no excessive 'sloppiness' or roughness, and check for any suspension noises when cornering and driving over bumps.

Check the performance of the engine, clutch (where applicable), gearbox and driveshafts

Listen for any unusual noises from the engine, clutch and gearbox.

Make sure that the engine runs smoothly when idling, and that there's no hesitation when accelerating.

Check that, where applicable, the clutch action is smooth and progressive, that the drive is taken up smoothly, and that the pedal travel is not excessive. Also listen for any noises when the clutch pedal is depressed.

On manual gearbox models, check that all gears can be engaged smoothly without noise, and that the gear lever action is not abnormally vague or 'notchy'.

On automatic transmission models, make sure that all gearchanges occur smoothly without snatching, and without an increase in engine speed between changes. Check that all the gear positions can be selected with the car at rest. If any problems are found, they should be referred to a Vauxhall dealer.

Listen for a metallic clicking sound from the front of the car, as the car is driven slowly in a circle with the steering on full lock. Carry out this check in both directions. If a clicking noise is heard, this indicates wear in a driveshaft joint, in which case seek advice and renew the joint if necessary.

Check the operation and performance of the braking system

Make sure that the car does not pull to one side when braking, and that the wheels do not lock prematurely when braking hard.

Check that there's no vibration through the steering when braking.

Check that the handbrake operates correctly, without excessive movement of the lever, and that it holds the car stationary on a slope.

SERVICING 119

Every 12 000 miles (20 000 km) or 1 year – whichever comes first, OR every 18 000 miles (30 000 km) or 2 years – whichever comes first

Note: *The shorter time/mileage interval shown above applies to models produced up to the end of September 1982. Models from October 1982 on use the longer interval for all these service tasks.*

In addition to all the items in the '6000/9000 miles (10 000/15 000 km)' service on page 106, carry out the following.

FUEL AND EXHAUST SYSTEMS

Renew the air cleaner filter element

Carburettor models
To remove the air cleaner element, remove the air cleaner cover. This is secured by a centre nut or bolt, or by three screws.

Additionally, release the spring clips around the edge of the cover or, if spring clips are not fitted, carefully prise around the lower edge of the cover with your fingers to release the retaining lugs.

With the cover removed, lift out the element and discard it.

▲ *Removing the air cleaner cover*

▲ *Removing the air cleaner filter element*

▲ *Removing the air cleaner centre screw*

▲ *Air cleaner cover alignment slot and lug*

VAUXHALL ASTRA & BELMONT

SERVICING

Wipe inside the air cleaner body, being careful not to introduce dirt into the carburettor throat. Remember to clean the inside of the top cover as well.

Fit the new element, then refit and secure the cover, observing any alignment marks or lugs.

Fuel injection models

On fuel injection models, the air cleaner housing is part of the airflow sensor housing. To remove the filter element, first release the locking clip, and disconnect the plug from the airflow sensor. Disconnect the air trunking.

Release the spring clips, and lift off the air cleaner cover with airflow sensor attached. Do not drop or jar the airflow sensor, as it is fragile (and expensive!)

▲ *Removing the air cleaner cover with airflow sensor attached*

▲ *Fitting the air cleaner filter element*

Remove the old element and discard it. Wipe clean the air cleaner body and cover. Fit the new element, making sure that its seal engages the cover recess.

Refit and secure the cover, then reconnect the airflow sensor plug. Refit the air trunking.

▲ *Fuel inlet union and gauze filter (typical)*

Clean the fuel inlet filter gauze (where fitted)

Disconnect the fuel inlet union from the carburettor, then unscrew the union plug and withdraw the gauze filter from the recess.

Clean the filter in fuel, then refit it and tighten the plug and union.

Check the exhaust system

This check is best carried out when the car is raised and supported on axle stands, as it will then be much easier to view the complete exhaust system. Look for black sooty stains, especially at joints, which indicate that exhaust gases are escaping. Also look for severe rusting, and check the condition of the exhaust mountings. Where rust or damage is evident, the exhaust system should be renewed or repaired, or the mountings renewed.

IGNITION SYSTEM

Renew the spark plugs

Refer to 'IGNITION SYSTEM' in the '6000/9000 miles (10 000/15 000 km)' service Section on page 112, and fit new spark plugs.

VAUXHALL ASTRA & BELMONT

SERVICING 121

Renew the contact breaker points

Refer to *'IGNITION SYSTEM'* in the *'6000/9000 miles (10 000/15 000 km)'* service Section on page 108, and fit a new set of contact breaker points.

CLUTCH

Check the clutch pedal adjustment

With progressive wear of the clutch linings, the pedal will over a period of time move upwards towards the steering wheel, making adjustment necessary.

Take a measurement from the edge of the steering wheel to the centre of the clutch pedal, with the pedal in its normal released position (dimension **A**), and then take another measurement with the pedal fully depressed (dimension **B**). These measurements can be taken using a suitable strip of wood or metal by making marks with the pedal released then depressed – the important thing is the *difference* between the two measurements.

The difference between the two measurements (B – A) is the pedal stroke, and this should ideally be 138.0 mm (5.4 in). If the pedal stroke measured is a long way off this figure, the cable needs adjusting. In the engine compartment, find the clutch release lever, which is in top of the gearbox, on the left-hand end of the engine (when viewed from the driver's seat).

▲ *Clutch pedal released measurement (A)*

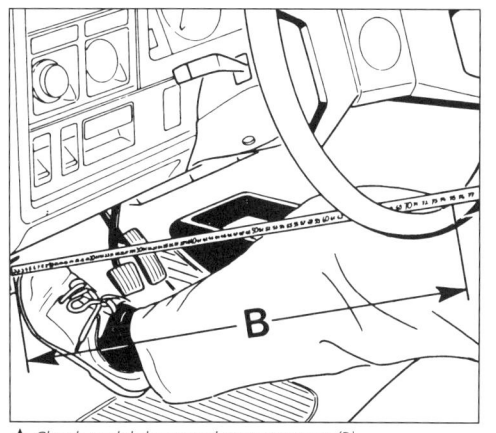

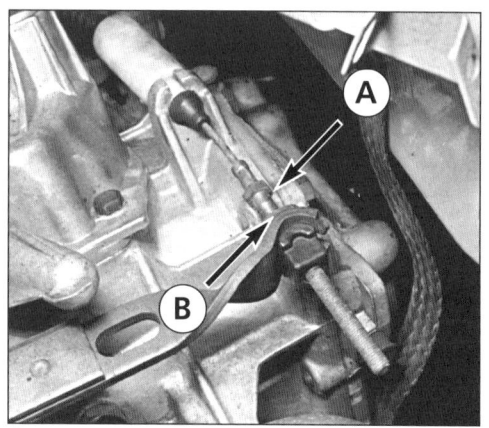

▲ *Clutch operating cable at the release lever on the gearbox*
A *Cable adjuster nut*
B *Spring clip*

The threaded end of the cable which fits into the clutch release lever must be adjusted to obtain the correct pedal movement. Remove the spring clip on the cable end fitting before making the adjustment. Turn the threaded end of the cable using the adjuster nut **(A)** until the correct pedal stroke is obtained.

Note that, when correctly adjusted, the clutch pedal will be higher than the brake pedal – if the clutch pedal is aligned with the brake pedal, the cable is in need of adjustment. Recheck the pedal movement after making an adjustment, and do not forget to refit the spring clip.

▲ *Clutch pedal depressed measurement (B)*

VAUXHALL ASTRA & BELMONT

SERVICING

BRAKING SYSTEM

Check the handbrake adjustment and linkages

It will only be necessary to adjust the handbrake cable if the cable has stretched, or if wear has occurred in the handbrake lever, since wear of the rear brake shoe linings is taken up by manual adjustment at the service interval on pre-October 1983 models, or by continual automatic adjustment on later models.

On pre-October 1983 models, ensure that the rear brake shoes are correctly adjusted (refer to the Owners Workshop Manual).

Chock the front wheels with blocks of wood or bricks, then jack up the rear of the car and support on axle stands. Apply the handbrake lever to its first notch, and check if the rear wheels can be rotated by hand – the brake shoes should be dragging slightly against the drums. The amount of drag should be the same on both rear wheels.

If adjustment is required, leave the handbrake lever on its first notch, then check that the cable compensator beneath the rear underbody is free to move. Using two spanners, adjust the cable by turning the self-locking nut on the threaded part of the cable end fitting on the compensator.

After making the adjustment, fully apply the handbrake lever (to the 4th or 5th notch), check that the rear wheels are locked, then release the lever and check that the wheels are free to turn. On models manufactured from October 1984 onwards, a further check may be made by removing the rubber plug on the inside of the brake backplate. When the handbrake adjustment is correct, the pin on the handbrake operating lever should be clear of the shoe web by approximately 3.0 mm (0.12 in) with the handbrake released.

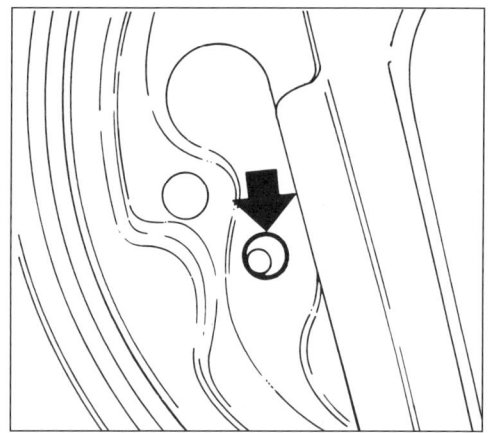

▲ *Handbrake lever pin (arrowed) visible through inspection hole on models manufactured from October 1984*

Check the handbrake cables for condition, making sure that there are no frayed strands or worn linkages. If the visible part of the inner cables is rusty, apply a little oil to prevent them seizing up. Check that the cables are mounted securely to the underbody.

Lower the car to the ground on completion.

SUSPENSION AND STEERING

Check power steering fluid level, and top-up if necessary (where applicable)

The power steering fluid level should be checked with the engine stopped. The dipstick is attached to the underside of the filler cap, on top of the fluid reservoir, which is located next to the battery.

When the fluid is at operating temperature (when the engine is hot), the level should be up to the FULL mark; when cold, the level should not be below the ADD mark.

Top-up if necessary using clean fluid of the specified grade (refer to '*Service specifications*'

▲ *Handbrake cable equaliser/adjuster – early models*

SERVICING 123

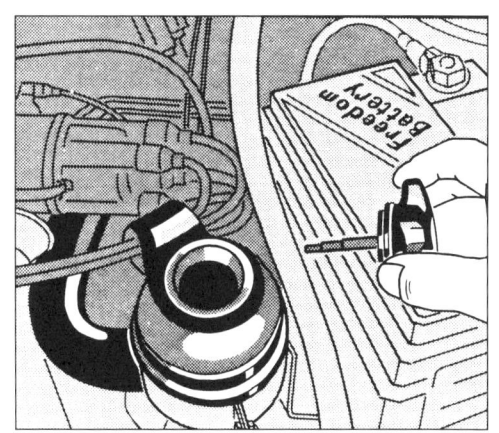

▲ *Power steering fluid filler cap and level dipstick*

on page 73). Look for leaks if regular topping-up is required.

ELECTRICAL SYSTEM

Check the condition and adjustment of the alternator drivebelt, and (where fitted) the power steering pump drivebelt

Refer to *'ELECTRICAL SYSTEM'* in the *'6000/9000 miles (10 000/15 000 km)'* service Section on page 116.

ROAD TEST

Refer to *'ROAD TEST'* in the *'6000/9000 miles (10 000/15 000 km)'* service Section on page 118.

ADDITIONAL TASKS

Note that as well as the tasks described in the preceding paragraphs, the additional tasks given in the *'Service schedule'* chart on pages 90 to 97 should also be carried out at this service interval. These tasks require more detailed explanation, or the use of special tools, and are considered beyond the scope of this Handbook. For details of these additional tasks, refer to the Owners Workshop Manual.

Every 2 years – regardless of mileage

COOLING SYSTEM

Renew the coolant

Note: *The following items will be required for this task:*
- *A suitable quantity of coolant solution (made up from approximately 55% clean water and 45% antifreeze)*
- *Suitable container to catch the old coolant as it drains – make sure the container capacity is greater than the volume of coolant in the system (refer to '**Service specifications**' on page 74)*
- *Screwdriver to disconnect the radiator hoses*

Draining the system

The coolant should be drained with the engine cold.

Remove the cooling system pressure cap. Refer to the *'Checking coolant level'* procedure on page 80 for details.

On 1.2 litre engines, there is a cylinder block drain plug on the side of the engine beneath the exhaust manifold, but the radiator bottom hose must be disconnected to drain the main system. On all engines except the 1.2 litre, there are no radiator or cylinder block drain plugs fitted, so the radiator bottom hose must be disconnected to drain the coolant.

Position a suitable container beneath the radiator bottom hose connection, then slacken the hose clip, release the hose, and allow the coolant to drain. Position the bottom hose as low down as possible, so that the maximum amount of coolant is drained.

On 1.2 litre engines, use an Allen key to unscrew and remove the cylinder block drain plug, and allow the coolant to drain. Note that a certain amount of coolant will remain in the cylinder block.

On models where the bottom hose outlet is not right at the bottom of the radiator (all automatic transmission models, and some manual gearbox models), note that some coolant will remain in the radiator after draining.

If rust or sludge is evident in the coolant which has been drained, the cooling system

SERVICING

should be flushed as described below to remove any further contamination from the engine and radiator.

If the coolant drained is clean, then the cooling system can be refilled as described later in this Section.

Flushing the system

To flush the radiator, disconnect the top hose from the radiator, then insert a garden hose into the top of the radiator. Allow the water to circulate until it runs clear from the bottom of the radiator. Due to the fact that the bottom hose outlet may not be at the bottom of the radiator on some types, it may be necessary to completely remove the radiator (refer to the Owners Workshop Manual) and turn it upside-down to be sure of cleaning away all sediment.

The thermostat must be removed in order to flush the engine, but this work is outside of the scope of this Handbook. For those wishing to carry out the work, refer to the Owners Workshop Manual for your car (OWM 635 for 'Mark 1' models, and OWM 1136 for 'Mark 2' and Belmont models).

Filling the system

To release air from the system as it is being filled, on 1.2 litre engines slacken the heater hose clip at the connection on the top of the engine.

On 1.3 & 1.4 litre engines, partially unscrew the coolant temperature sender from the inlet manifold (the inlet manifold is shown on all the underbonnet component location illustrations earlier in this Chapter).

On 1.6, 1.8 & 2.0 litre engines, loosen the bleed screw or coolant temperature sender on the thermostat housing (the thermostat housing is shown on the underbonnet component location illustrations earlier in this Chapter). An Allen key will be required for this on the 2.0 litre 16-valve engine.

Before refilling, make sure that all hoses are re-connected (do not over-tighten the hose clips, or you'll damage the rubber hoses), and on 1.2 litre engines, refit and tighten the cylinder block drain plug. A solution of 55% clean water and 45% antifreeze should be used to fill the system all year round. It's important to note that because of the different types of metals used in the engine, it's vital to use antifreeze with suitable anti-corrosion additives all year round. **Never** use water alone to fill the whole system.

Fill the system slowly through the expansion tank filler. Keep filling the system until coolant free of bubbles emerges at the heater hose, bleed screw or temperature sender (as applicable), then tighten the hose clip, bleed screw or temperature sender. Continue filling the system until the coolant is slightly above the KALT (cold) mark on the expansion tank. Squeezing the top and bottom hoses will help to dislodge air pockets in the system.

Screw the expansion tank filler cap on tight, then run the engine at a fast tickover, watching for leaks of coolant, until the electric cooling fan cuts in. Stop the engine and allow it to cool for at least two hours, then if necessary top-up the expansion tank so that the coolant is a little

▲ *Heater hose outlet on the OHV engine*

▲ *Coolant temperature sender location on the 1.3 litre engine*

SERVICING

above the KALT (cold) mark. Remember that the engine must be cold in order to obtain an accurate level, as it is quite normal for the coolant level to appear too high when the engine is hot. Refit and tighten the expansion tank filler cap on completion.

Every 24 000 miles (40 000 km) or 2 years – whichever comes first

In addition to the items listed in the previous services, the additional automatic transmission task given in the *'Service schedule'* chart on pages 90 to 97 should be carried out every 24 000 miles (40 000 km) or 2 years – whichever comes first. For details, refer to the Owners Workshop Manual.

Every 36 000 miles (60 000 km) or 4 years – whichever comes first

In addition to the items listed in the previous services, the additional engine and automatic transmission tasks given in the 'Service schedule' chart on pages 90 to 97 should be carried out every 36 000 miles (60 000 km) or 4 years – whichever comes first. For details, refer to the Owners Workshop Manual.

Every 74 000 miles (120 000 km) or 8 years – whichever comes first

In addition to the items listed in the previous services, the additional engine task given in the *'Service schedule'* chart on pages 90 to 97 should be carried out every 74 000 miles (120 000 km) or 8 years – whichever comes first. For details, refer to the Owners Workshop Manual.

SEASONAL SERVICING

If you carry out all the procedures described in the previous Section, at the recommended mileage or time intervals, then you'll have gone a long way towards getting the best out of your car in terms of both performance and long life. In spite of this, there are always other areas, not dealt with in regular servicing, where neglect can spell trouble.

A little extra time spent on your car at the beginning and end of every Winter will be well worthwhile in terms of peace of mind and prevention of trouble. The suggested tasks which follow have therefore been divided into Spring and Autumn Sections – but it's always a good idea to do them more frequently if you feel able.

Autumn

COOLING SYSTEM

Check the radiator and all hoses for signs of deterioration or damage

Refer to *'ENGINE COMPARTMENT'* in the *'6000/9000 miles (10 000/15 000 km)'* service Section on page 106.

Check the strength of the coolant solution

Refer to *'COOLING SYSTEM'* in the *'6000/9000 miles (10 000/15 000 km)'* service Section on page 107.

SERVICING

FUEL AND EXHAUST SYSTEM

Adjust air cleaner setting (1.2 litre engine only)

The lever on the side of the air cleaner housing should be adjusted according to the outside air temperature as follows. The temperatures are given on the side of the air cleaner.

▲ *Air cleaner intake air temperature adjustment lever*
SUMMER *Above 10°C*
INTERMEDIATE *10°C to -5°C*
WINTER *Below -5°C*

A check should be made of the setting throughout the Autumn, Winter and Spring period, to ensure that it corresponds with the prevailing outside temperature.

ELECTRICAL SYSTEM

Where applicable, check the battery electrolyte level

Refer to *'Regular checks'* on page 84.

Check and if necessary clean the battery terminals

Make sure that the terminals are secure, and clean any corrosion from the metal. Be sure to wear eye protection for this, as the white-coloured deposits are harmful. To prevent corrosion, the terminals can be coated with petroleum jelly (don't use ordinary grease). It's worthwhile examining the battery tray at the same time, and if the same white-coloured deposits are present, clean them using a brush (or wash them away with plenty of hot water, being careful to avoid splashes).

Check the condition and adjustment of the alternator drivebelt

Refer to the procedure given in the *'6000/9000 miles (10 000/15 000 km)'* service Section on page 116.

Check the operation of all lights, electrical equipment and accessories

Refer to *'Fault finding'* on page 143 if any problems are discovered.
   If any of the bulbs need to be renewed, refer to *'Bulb, fuse and relay renewal'* on page 131.

Check the washer fluid level, and the wipers and washers

Refer to *'Regular checks'* on page 85.

TYRES

Check tread depth and condition

Refer to *'Regular checks'* on page 82. Remember that you may be driving in slippery conditions during the Winter.

BODYWORK

Thoroughly clean the car

Wash the car, and then polish it thoroughly to help protect the paint during the Winter.

Spring

UNDERBODY

Thoroughly clean the underside of the car

The best time to clean the underside of the car is after the car has been driven in wet conditions, when the accumulated dirt will be softened up.
   To clean the car, first of all the car must be jacked up as high as possible (making sure that it is safely supported – refer to *'Jacking and vehicle support'* in *'Safety first!'* on page 100).
   Gather together a quantity of paraffin, or water-soluble solvent, a stiff-bristle brush, a scraper, and a garden hose.
   With the car jacked up and safely supported,

SERVICING

get underneath the car, and cover the brake components at each wheel with polythene bags, to stop dirt and water getting into them.

Loosen any encrusted dirt, scraping or brushing it away – the paraffin or solvent can be used where there's oil contamination. Pay particular attention to the wheel arches. Take care not to remove the underseal (a waxy substance applied to the car underbody to protect against corrosion).

When all the dirt is loosened, a wash down with the hose will remove the remaining dirt and mud.

You can now check for signs of damage to the underseal. If there's any sign of the underseal breaking away, patch it up by spraying or brushing on a suitable underseal wax (available from motor accessory shops and motor factors). Make sure the area is clean and dry before applying the wax.

Take the opportunity to check for signs of rusting. Likely places are the body sills and floor panels.

If rust is found, seek advice and have the affected area repaired before the problem gets too bad.

On completion, lower the car to the ground.

BODYWORK

Thoroughly check and clean the surfaces of the bodywork

Give the car a thorough wash, and check for stone chips and rust spots. For details of how to treat these minor paint problems, refer to 'Bodywork and interior care' on page 130.

After any repairs to the paint have been carried out (not *before*, otherwise the paint will not stick), give the car a polish.

VAUXHALL ASTRA & BELMONT

TOOLS

WHAT TO BUY

If you're intending to carry out your own servicing, you'll need to obtain a few basic tools. Although at first sight you may think that tools seem expensive, once you've bought them they should last a lifetime if you look after them properly.

The tools supplied with the car will enable you to change a wheel, and not much more. The absolute minimum tool kit you'll need to carry out any maintenance or servicing will be a range of metric spanners, two screwdrivers (one for crosshead or 'Phillips' type screws), and a pair of pliers. With a bit of ingenuity, these items should enable you to complete the more basic routine servicing jobs, but they won't allow you to do much else.

When buying tools, it's important to bear in mind the quality. You don't need to buy the most expensive tools available, but generally you get what you pay for. Cheap tools may prove to be a false economy, as they're unlikely to last as long as better quality alternatives.

It's very difficult to lay down hard and fast rules on exactly what you're going to need, but the following list should be helpful in building up a good tool kit. Combination spanners (ring one end, open-ended the other) are recommended because, although more expensive than double open-ended ones, they give the advantages of both types.

- Combination spanners to cover a reasonable range of sizes (say 7 mm to 17 mm at least)
- Adjustable spanner
- Spark plug spanner (with rubber insert)
- Spark plug gap adjustment tool
- Set of feeler gauges
- Screwdriver (Plain) – 100 mm long blade x 6 mm diameter (approx)
- Screwdriver (Crosshead) – 100 mm long blade x 6 mm diameter (approx)
- Oil filter wrench
- Pliers
- Tyre pump
- Tyre pressure gauge
- A suitable key (Torx or Allen type, depending on model) to fit the gearbox oil filler plug (later models only)
- Oil can
- Funnel (medium size)
- Stiff brush (for general cleaning jobs)
- Tool box
- Hydraulic jack (ideally a trolley jack)
- Pair of axle stands
- Suitable containers for draining engine oil and coolant
- Inspection lamp

This is by no means a comprehensive list of tools, and you'll probably want to gradually add to your tool box as you discover the need for other tools, especially if you decide to tackle some of the more advanced servicing jobs described in the **Owners Workshop Manual** for your particular car.

CARE OF YOUR TOOLS

Having bought a reasonable set of tools, it's worth taking the trouble to look after them. After use, always wipe off any dirt or grease using a clean, dry cloth before putting them away. Never leave tools lying around after they've been used – a tool rack, or better still a proper toolbox will prove the best way of keeping everything up together. Rags can be wrapped around loose tools to prevent them from rattling if you're going to keep them in the boot of the car.

Feeler gauges should be wiped with an oily cloth from time to time, to leave a thin coating of oil on the metal which will prevent corrosion. Screwdriver blades inevitably lose their keen edges, and a little occasional attention with a file or an oilstone will keep them in good condition.

BODYWORK AND INTERIOR CARE

It's well worth spending the time and effort necessary to look after your car's bodywork and interior. Cleaning your car regularly will not only improve its appearance, it will also protect the bodywork against the elements and the grime encountered in everyday driving. It's worth bearing in mind that if your car looks clean and tidy, it will also be worth more money when you come to sell it or trade it in for a newer model.

This Section will help you to keep your car in 'showroom' condition. If you wish to tackle repairs to more serious bodywork damage or corrosion, comprehensive details can be found in our **'Car Bodywork Repair Manual'**.

CLEANING THE INTERIOR

It's a good idea to clean the interior of the car first, before cleaning the exterior, as this will avoid spreading the dirt from inside over the bodywork.

Start by removing all the loose odds and ends from inside the car, not forgetting the ashtrays, glovebox and any interior pockets. Take out any loose mats or carpets, which should be shaken and brushed, and if possible vacuum-cleaned.

The inside of the car can be cleaned with a brush and dustpan, or preferably a vacuum-cleaner. If you can't use your household vacuum-cleaner, it may be worth investing in one of the small 12-volt hand vacuum-cleaners which can be powered from the car battery. If the carpets are very dirty, use a suitable proprietary cleaner with a brush to remove the ingrained dirt. Ideally, it's best to remove the carpets for cleaning, but this is an involved task in most modern cars, and it will probably be easier to leave them in place.

The facia, door trim, seats, and any other items which require attention can now be wiped over with a cloth soaked in warm water containing a little washing up liquid. If the trim is particularly dirty, use one of the proprietary cleaners available from motor accessory shops – a number of different types are available, suitable for vinyl, cloth or leather upholstery, etc, as required. An old nail brush or tooth brush will help to remove any ingrained dirt. On completion, wipe the surfaces dry using a lint-free cloth, and leave the windows open to speed up drying.

The inside of the windscreen and windows should be cleaned using a proprietary glass cleaner, or methylated spirits. Be careful about using certain household products such as washing-up liquid which may leave a smeary film. Finish off by wiping with a clean, dry paper tissue.

Don't forget to clean the boot.

Check for any nicks or tears in the interior trim, seats, headlining, etc. Various repair kits are available from motor accessory shops to suit most types of trim, but if the damage is very serious, the relevant trim panel will probably have to be renewed.

CLEANING THE BODYWORK

Ideally, the car should be washed every week, either by hand (preferably using a hosepipe), or by using a local car-wash. If you're washing the car by hand, use plenty of water to loosen the dirt and dust, and if possible use a suitable car shampoo or wax additive in the water. Any stubborn dirt such as road tar or bird droppings can be removed using methylated spirit or preferably one of the special proprietary cleaning solutions (in which case make sure that the product is suitable for use on car paint) – in either case, wash the affected area down with plenty of water after removing the dirt. **Never** just wipe over a dirty car, as this will scratch the paint very effectively.

VAUXHALL ASTRA & BELMONT

BODYWORK AND INTERIOR CARE

Two or three times a year, a good silicone or wax polish can be used on the paintwork. It's important to wash the car and remove all stubborn dirt, tar, etc, before using polish, as the polish will effectively seal over the top of any dirt, making it extremely difficult to remove in the future. Good polishes actually form a protective coating over the paint finish, which should help to make future accumulated dirt easier to remove. Always follow the manufacturer's recommendations closely when using polish. Try to avoid getting polish on any unpainted plastic or rubber body parts such as bumpers and spoilers, as it tends to leave a stain when dry. Chrome parts are best cleaned with a special chrome cleaner, as ordinary metal polish will wear away the finish.

Plastic or rubber body parts such as bumpers and spoilers should be cleaned using a suitable proprietary cleaner. A number of different types are available, including colour restorers, and special solvents to remove any stray polish which may have crept onto the plastic when polishing the paintwork. Make absolutely sure that any solvents or cleaners used are suitable, as certain products may attack plastic and/or rubber.

If the paint is beginning to lose its gloss or colour, and ordinary polishing doesn't seem to solve the problem, it's worth considering the use of a polish with a mild 'cutting' action to remove what is in effect a surface layer of 'dead' paint. In this case, follow the manufacturer's instructions, and don't polish too vigorously, or you might remove more paint than you intended!

DEALING WITH SCRATCHES

With superficial scratches which don't penetrate down to the metal, repair can be very simple.

Very light scratches can be polished out by carefully using a suitable 'cutting' polish. Follow the manufacturer's instructions, and take care not to remove too much of the surrounding paint.

If the scratch cannot be polished out, touch-up paint will be required. Touch-up paint is usually available in the form of a touch-up stick from the car manufacturer's dealers, or in various forms from motor accessory shops. You will probably have to quote the year and model of the car in order to make sure that you obtain the correct matching colour, and in some cases you may have to quote a paint reference number which will usually appear on the car's Vehicle Identification Number (VIN) plate under the bonnet (refer to *Servicing* on pages 101 to 104).

Lightly rub the area of the scratch with a very fine cutting paste to remove loose paint from the scratch and to clear the surrounding bodywork of polish.

Rinse the area with clean water.

Apply suitable touch-up paint to the scratch using a fine paint brush, and continue to apply fine layers of paint until the surface of the paint in the scratch is level with the surrounding paintwork.

Allow the new paint at least two weeks to harden, then blend it into the surrounding paintwork by rubbing the scratch area with a paintwork renovator or a very fine cutting paste.

If the paint finish requires a lacquer coat (in which case the lacquer will be supplied with the touch-up paint), it should now be applied to the newly painted area and allowed to dry in accordance with the manufacturer's instructions.

Finally, apply a suitable wax polish to the affected area for added protection.

VAUXHALL ASTRA & BELMONT

BULB, FUSE AND RELAY RENEWAL

BULBS

A defective exterior light can be not only dangerous, but also illegal. Carrying spare bulbs will enable you to renew blown ones as they occur. A failed interior light bulb may be just a nuisance, but a faulty exterior light could be a life-or-death matter.

Before assuming that a bulb has failed, check for corrosion on the bulbholder and wiring connections, particularly around the rear light assemblies.

- Note that as a safety precaution, the battery earth (black) lead should always be disconnected before renewing a bulb.
- Remember to reconnect the lead on completion.

Bulb ratings

Always make sure that when a bulb is renewed, a new bulb of the correct rating is used. All the bulbs are of the 12-volt type, and their ratings are as follows:

Bulb	Rating [watts]
Headlight (halogen)	60/55
Headlight (standard)	45/40
Foglight (front)	55
Spotlight	55
Direction indicators, front and rear	21
Foglight (rear)	21
Reversing light	21
Brake and tail light	21/5
Rear number plate light	10
Interior courtesy light	10
Boot light	10
Engine compartment light	10
Glove compartment light ('Mark 1' models)	5
Glove compartment light ('Mark 2' and Belmont models)	10
Side (parking) light ('Mark 1' models)	4
Side (parking) light ('Mark 2' and Belmont models)	5
Side repeater light	5
Alternator warning light	3
Instrument lighting	1.2
Warning lights	1.2
Switch illumination	1.2
Cigarette lighter and ashtray lights ('Mark 1' models)	1.2
Cigarette lighter and ashtray lights ('Mark 2' and Belmont models)	0.5

132 BULB, FUSE AND RELAY RENEWAL

Exterior light bulbs

HEADLIGHT

Working in the engine compartment, disconnect the wiring plug from the rear of the headlight.

▲ *Disconnecting the headlight wiring plug*

Remove the rubber cover to expose the retaining ring or spring clip which secures the bulb.

▲ *Rubber cover removed to expose the spring clip*

Turn the retaining ring anti-clockwise, or release the spring clip, then withdraw the bulb from the rear of the headlight and discard it; remember that it may be very hot if it has just been in use.

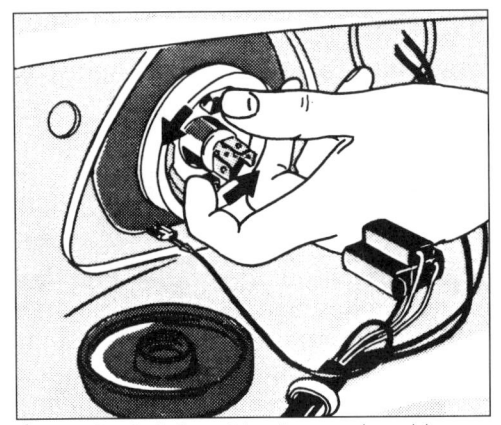

▲ *Removing the bulb retaining ring on early models*

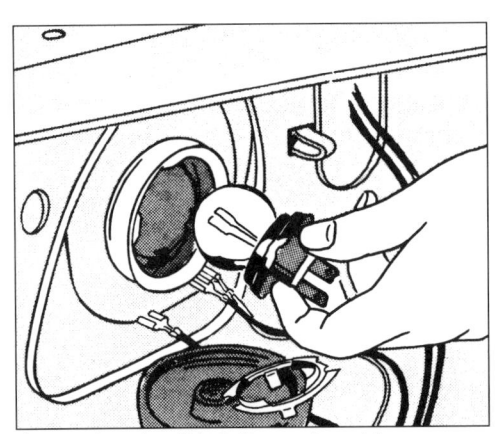

▲ *Removing a headlight bulb on early models*

▲ *Removing a headlight bulb on later models*

VAUXHALL ASTRA & BELMONT

BULB, FUSE AND RELAY RENEWAL 133

Fit the new bulb, being careful not to touch the glass of the bulb with the fingers. Clean the glass with meths or white spirit if the glass is accidentally touched, otherwise the grease from your fingertips may cause blackening and premature failure of the bulb. Make sure that the lugs on the bulb engage in the recesses in the bulbholder.

FRONT SIDELIGHT

Working in the engine compartment, remove the sidelight bulbholder from the headlight reflector by depressing it and turning it anti-clockwise. Depress and twist the bulb to remove it from the bulbholder.

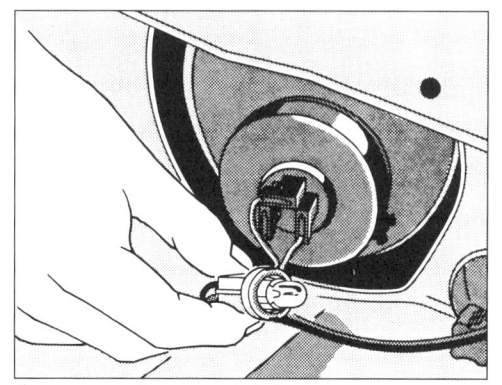

▲ *Removing the front direction indicator light bulbholder on 'Mark 1' models*

▲ *Removing a front sidelight bulb*

Fit the new bulb, making sure that it is correctly engaged with the bulbholder, then insert the bulbholder into the headlight reflector and turn clockwise to lock.

▲ *Removing the front direction indicator light bulbholder on 'Mark 2' and Belmont models*

FRONT DIRECTION INDICATOR LIGHT

Working in the engine compartment, on 'Mark 1' models, remove the bulbholder by twisting it and pulling it out of the light unit.

On 'Mark 2' and Belmont models, remove the bulbholder by squeezing its legs together and twisting it anti-clockwise.

Remove the bulb from the bulbholder by depressing it and twisting it anti-clockwise.

Fit the new bulb to the bulbholder, and refit the bulbholder to the light unit.

FRONT FOGLIGHT ('MARK 1' MODELS)

Remove the cross-head screw from the bottom of the foglight, and carefully pull the lens/reflector unit out of the housing.

Release the spring clips which secure the bulb retainer, extract the bulb, then disconnect the wire leading to the bulb at the plug connector.

Fit the new bulb, being careful not to touch the glass with the fingers. Note how the recesses in the bulb flange mate with the pips on the reflector. Reconnect the wire, secure the bulb retainer and refit the lens/reflector unit to the housing.

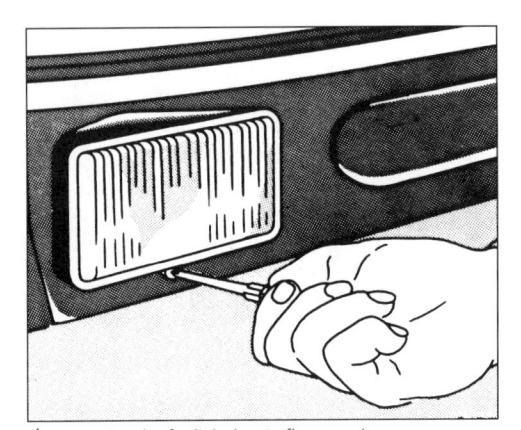

134 BULB, FUSE AND RELAY RENEWAL

FRONT FOGLIGHT ('MARK 2' AND BELMONT MODELS)

Turn the front wheels fully to the left or right to gain access to the relevant front foglight located in the front of the wheelarch. Remove the rear cover from the foglight by twisting it anti-clockwise.

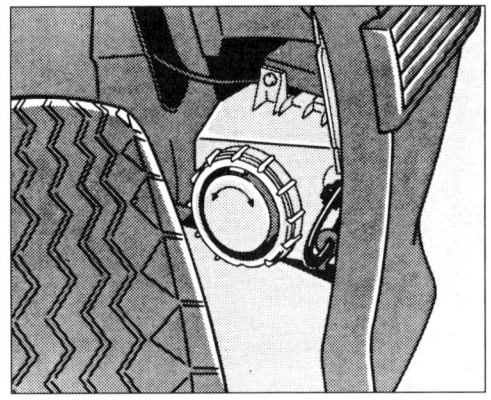

▲ *Removing the foglight lens/reflector unit retaining screw*

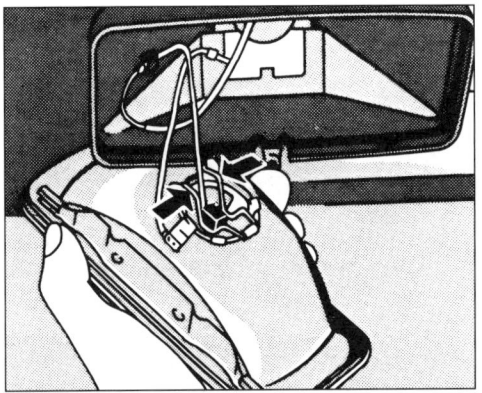

▲ *Rear cover on the foglight*

Release the spring clips, then remove the bulb from the reflector unit. Disconnect the wire leading to the bulb at the plug connector.

Fit the new bulb, being careful not to touch the glass with the fingers. Make sure that the cut-outs in the bulb flange engage with the pips on the reflector unit.

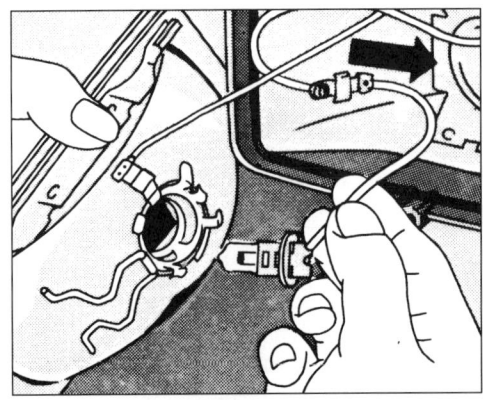

▲ *Push spring clip legs in direction arrowed to remove the front foglight bulb*

▲ *Removing the front foglight bulb – arrow indicates separation of wire connector*

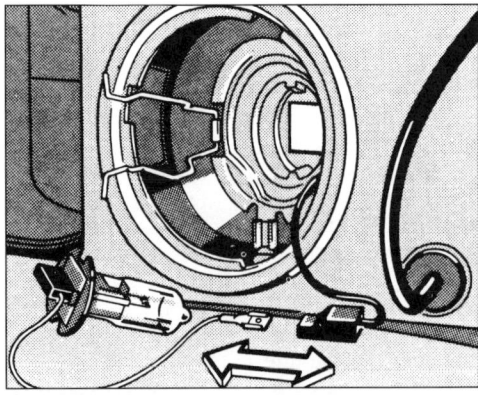

▲ *Removing the foglight bulb and separating the wire connector*

VAUXHALL ASTRA & BELMONT

BULB, FUSE AND RELAY RENEWAL 135

FRONT DIRECTION INDICATOR REPEATER LIGHT

Turn the lens anti-clockwise to release it from the bayonet fitting. Pull out the bulb – it is of the wedge-base type.

▲ Front direction indicator repeater light bulb renewall

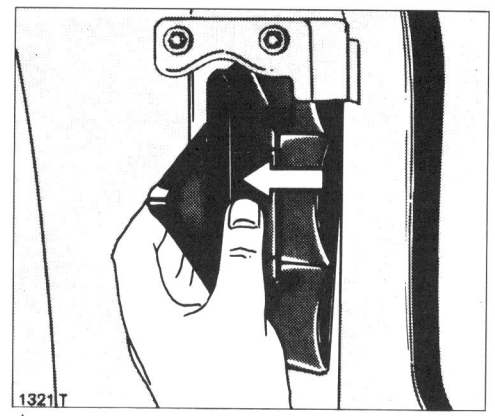

▲ Press in the direction of the arrow to release the bulbholder

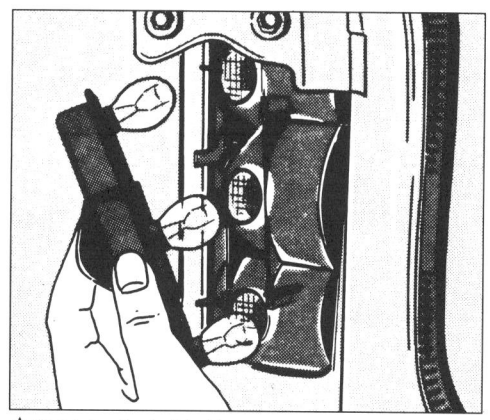

Press the new bulb firmly into the bulbholder, then refit the lens and tighten clockwise.

REAR LIGHT CLUSTER

'Mark 1' models

Prise out the cover in the luggage compartment, then remove the bulbholder by pressing it outwards.

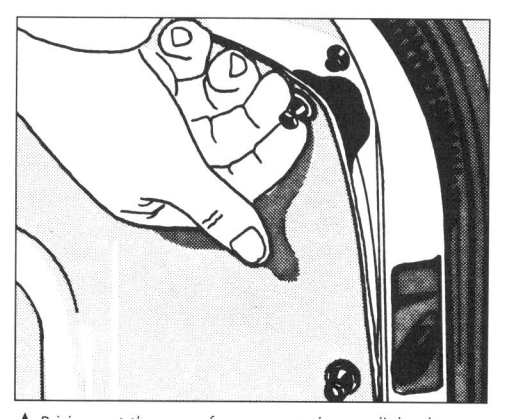

▲ Removing the bulbholder

'Mark 2' and Belmont models

On Astra Hatchback and Estate models, depress the lock tongue or twist the knob to open the cover flap, then remove the bulbholder by depressing the clip on the top and slightly raising the bulbholder.

On Belmont models, prise out the cover from the rear corner of the luggage compartment, then depress the clip on the top of the bulbholder and slightly raise it to remove it.

With the bulbholder removed, depress and twist the appropriate bulb to remove it. Fit the new bulb, then refit the bulbholder and cover.

▲ Prising out the cover for access to the rear light cluster

VAUXHALL ASTRA & BELMONT

136 BULB, FUSE AND RELAY RENEWAL

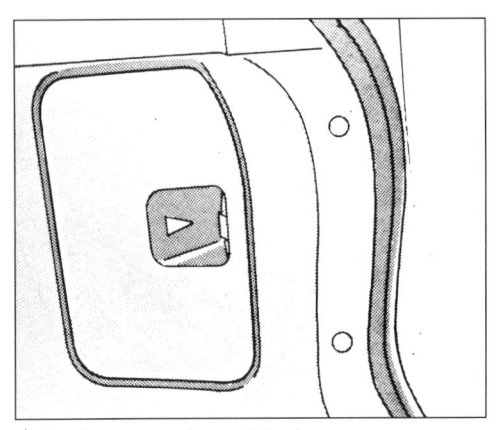

▲ *Flap for access to the rear light cluster*

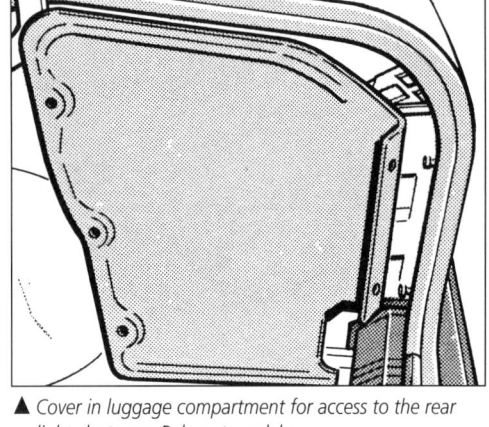

▲ *Cover in luggage compartment for access to the rear light cluster on Belmont models*

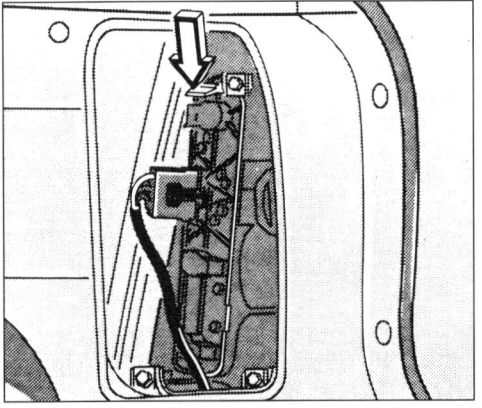

▲ *Press in the direction of the arrow to release the bulbholder*

REAR NUMBER PLATE LIGHT

Press the number plate light to the left slightly, then use a small screwdriver to prise the right-hand side of the light out of the bumper.

Separate the base of the light from the lens by depressing the protruding tongue – on Estate models, there are two tongues. Depress and twist the bulb to remove it.

Fit the new bulb, assemble the lens to the base, then press the light unit into the rear bumper until it clicks firmly into position.

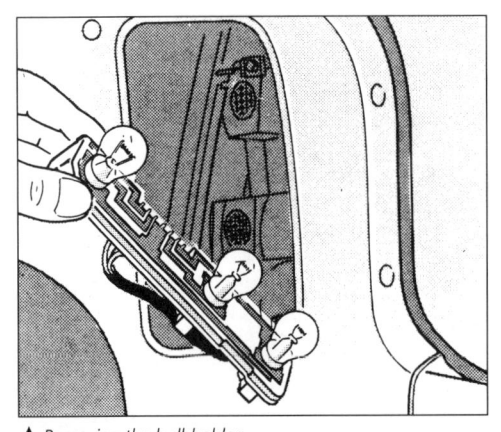

▲ *Removing the bulbholder*

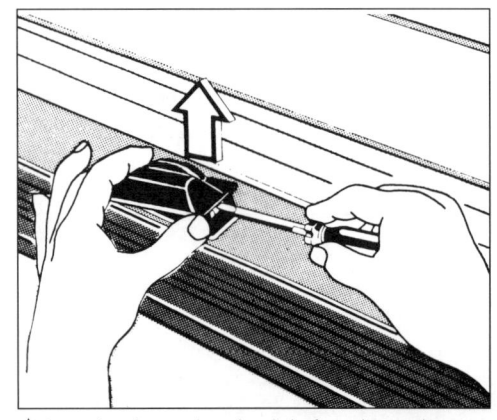

▲ *Removing the number plate light from the rear bumper*

VAUXHALL ASTRA & BELMONT

BULB, FUSE AND RELAY RENEWAL   137

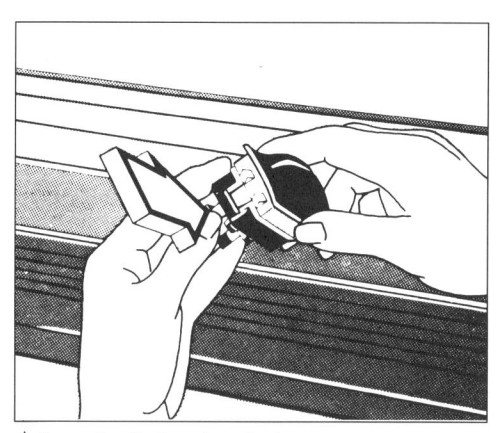

▲ *Separating the base from the lens*

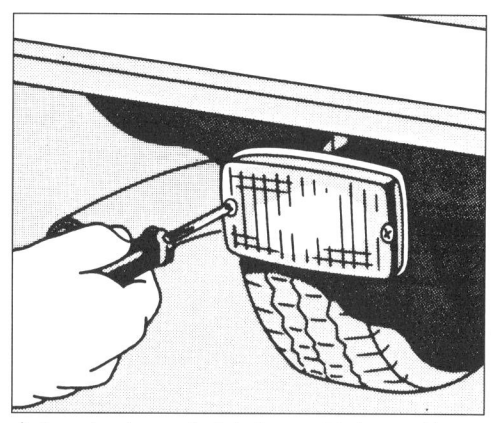

▲ *Removing the rear foglight lens on 'Mark 1' models*

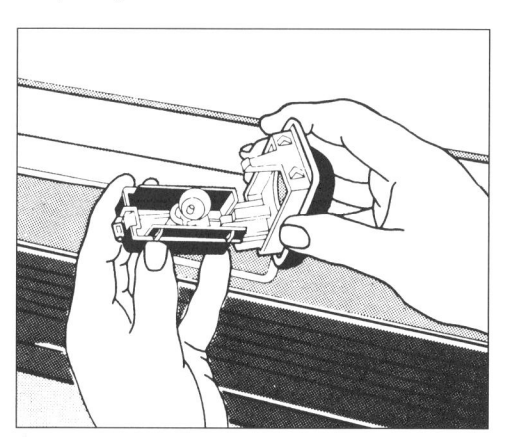

▲ *The bayonet type bulb is located in the base*

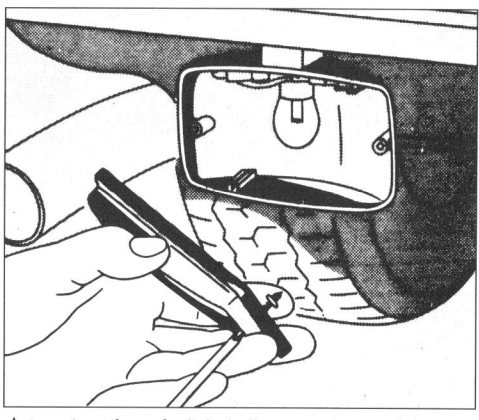

▲ *Location of rear foglight bulb on 'Mark 1' models*

REAR FOGLIGHT

On 'Mark 1' models, the rear foglight is a separate unit located below the rear bumper; on 'Mark 2' and Belmont models it is incorporated in the rear bumper. To remove the bulb, first unscrew the two cross-head screws and remove the lens. Depress the bulb into its holder, and twist it to remove.

Fit the new bulb, then refit the lens and gently tighten the screws.

REVERSING LIGHT

On all models, the reversing light is the bottom bulb in the rear light cluster, and its removal is described under '*REAR LIGHT CLUSTER*' on page 135.

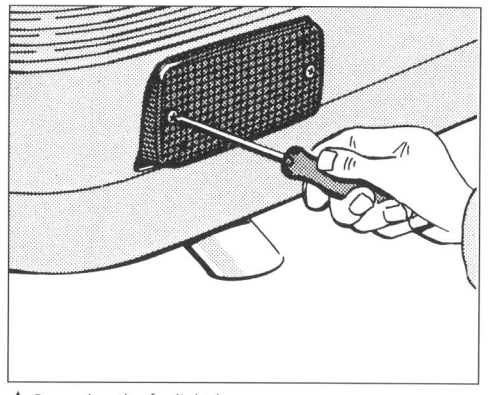

▲ *Removing the foglight lens on 'Mark 2' and Belmont models*

VAUXHALL ASTRA & BELMONT

BULB, FUSE AND RELAY RENEWAL

Interior light bulbs

Note: *Renewal of the various instrument panel and facia-mounted control illumination light bulbs requires detailed explanation, and in some cases extensive dismantling. Details can be found in the relevant Owners Workshop Manual for your particular car (OWM 635 for 'Mark 1' models, or OWM 1136 for 'Mark 2' and Belmont models)*

COURTESY LIGHT

On models with a one-piece light unit, carefully prise the courtesy light from its location using a small screwdriver (taking care not to damage the surrounding trim), then pull the bulb from the spring contacts in the light body. On some models, it is possible to remove the lens separately, leaving the main body of the light in position.

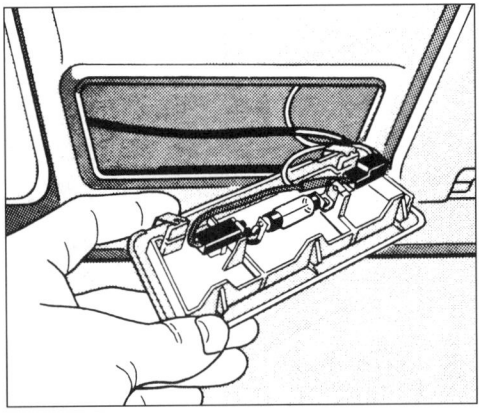

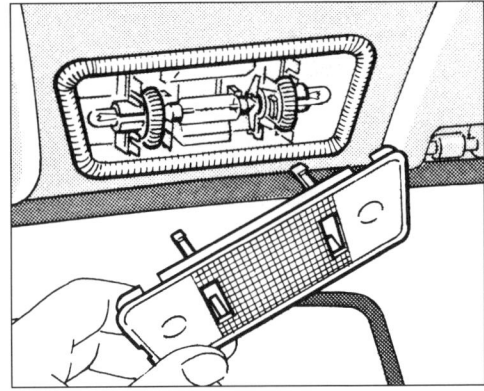

▲ *Removing the lens*

Fit the new bulb, making sure that it locates securely in the spring contacts, then refit the light or lens by pressing it lightly into its location.

ENGINE COMPARTMENT/LUGGAGE COMPARTMENT LIGHT

Using a small screwdriver, carefully prise the light assembly from its location, taking care not to damage the surrounding trim, then pull the bulb from the spring contacts in the light body.

Fit the new bulb, making sure that it locates securely in the spring contacts, then refit the light assembly.

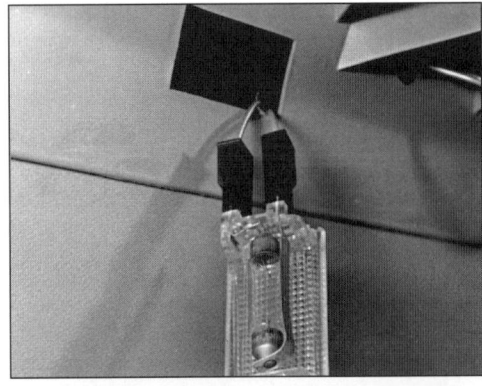

▲ *Removing the courtesy light located at the front of the headlining (one-piece type light unit)*

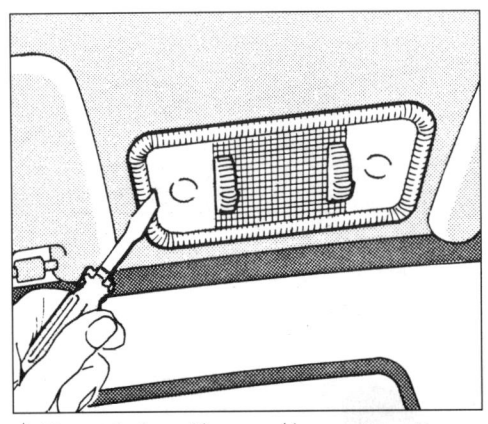

▲ *Prise out the lens with a screwdriver*

▲ *Luggage compartment courtesy light removed from its location*

VAUXHALL ASTRA & BELMONT

BULB, FUSE AND RELAY RENEWAL

FUSES

Most of the car's electrical circuits are protected by fuses, which will blow when the relevant circuit becomes overloaded. This is to prevent possible damage to electrical components, or the possible risk of fire.

- Before renewing a fuse, always make sure that the ignition and the relevant circuit are switched off.

The fuses are located in a fusebox mounted to the right of the steering column, on the lower part of the facia. Access to the fuses is gained by removing the cover.

The fusebox also contains the various relays.

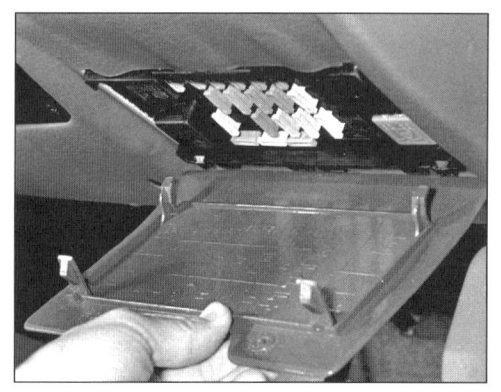

On models fitted with a headlamp washer system or folding roof, the relevant fuse(s) and relay(s) are mounted under the bonnet on the left-hand suspension turret.

The fuses for the central locking and electric windows are located on the back of the fusebox, and are accessed by unclipping the fusebox.

All fuses are a push fit in their sockets, and a blown fuse can be recognised by a break in the wire connection between the two terminals.

If a fuse has blown, renew it with one of identical rating. The fuses are colour-coded to show their rating as follows:

Colour	Rating (amps)
Red	10
Blue	15
Yellow	20
Green	30

There are no spare fuses in the fusebox, so it is recommended that a spare set of fuses is obtained from a Vauxhall dealer or motor factors, and carried in the car.

Always use proper fuses of the correct type and rating, and **never** use wire or any other material to bridge the gap where a fuse should be.

Note that the same fuse should never be renewed more than once without investigating the source of the trouble. Seek advice from someone suitably qualified if necessary.

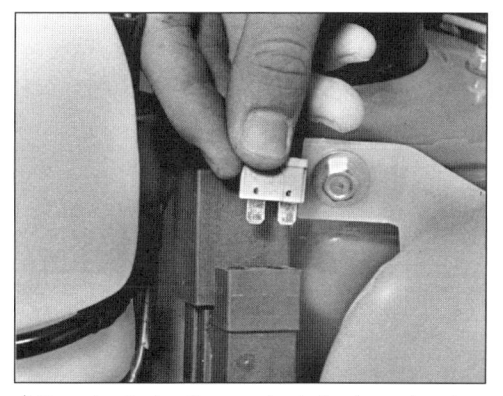

▲ Removing the fusebox cover

▲ Removing the headlamp washer in-line fuse – the relay is located behind the fuse

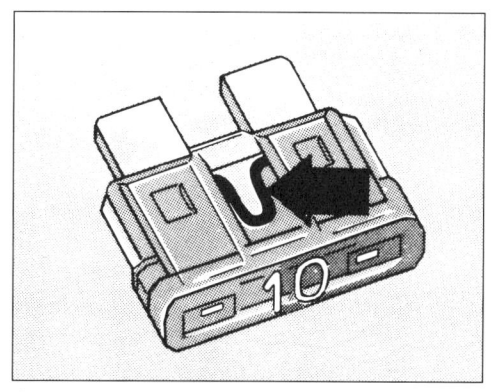

▲ A 10 amp fuse, showing the fuse wire (arrowed) which will melt when the circuit is overloaded

VAUXHALL ASTRA & BELMONT

140 BULB, FUSE AND RELAY RENEWAL

Various additional in-line fuses may be used, depending on any additional equipment fitted. If a problem is suspected which may be due to a fuse not located in the fusebox, it's best to seek advice from a Vauxhall dealer or auto electrician who will know where to look.

RELAYS

Various types of relays are used in some circuits, and a faulty relay may prevent one or more components from working.

- **Before renewing a relay, always make sure that the ignition and the relevant circuit are switched off.**

The main relays are located in the fusebox, and are of the plug-in type. Four of the relays are accessible from the outer face of the fusebox, but some (according to model) are accessible from the inner side of the fusebox by depressing the upper detent then unclipping the fusebox from the facia. Other relays may be mounted under the bonnet on the left-hand suspension turret. To remove a relay, simply pull it out of its socket.

It's difficult to tell visually whether a relay is faulty, and the best test is to substitute a known good relay to see if the relevant circuit then operates – if it does, then it's probably the relay which was causing the problem.

▲ *Relays on the outer face of the fusebox*
  A *Flasher unit*
  B *Heated rear window relay*
  C *Windscreen wiper relay/delay unit*
  D *Front foglight relay*

▲ *Relays on the inner face of the fusebox*
  E *Trip computer relay (not UK)*
  F *Rear wiper relay*
  G *Day running lamp relay (not UK)*
  H *Headlamp cut-out (not UK)*
  I *Headlamp warning buzzer*
  19 *Central locking fuse*
  20 *Electric windows fuse*

VAUXHALL ASTRA & BELMONT

PREPARING FOR THE MOT TEST

THE MOT TEST

The following information is intended as a guide to enable you to spot some of the more obvious faults which may cause your car to fail the MOT test.

Obviously it isn't possible to check the car to the same standard as a professional MOT tester, who will be highly experienced and will have all the necessary tools and equipment.

Although we can't cover all the points of the test, the following should provide you with a good indication as to the general condition of the car, and will enable you to identify any obvious problem areas before submitting your car for the test.

Further explanation of most of the checks, and details of how to cure any problems discovered can be found in the relevant *Owners Workshop Manual* for your particular car (OWM 635 Astra 1980 to October 1984, or OWM 1136 Astra & Belmont October 1984 to October 1991).

Lights
- All lights must work reliably, and the lenses and reflectors must not be damaged.
- Pairs of similar lights must be of the same brightness (eg both rear lights).
- Both headlights must show the same colour, and must be correctly aimed so that they light up the road adequately, but don't dazzle other drivers.
- The brake lights must work when the brake pedal is pressed.
- The direction indicator lights must flash between one and two times per second.
- All switches and driver's tell-tale lights must be fitted, secure and in good working order.
- No lamp should be adversely affected by the operation of any other lamp.

Steering and suspension
- All the components and their mounting points must be secure, and there must be no damage or excessive corrosion.
- There must be no excessive free play in the joints and bushes, and all rubber gaiters must be secure and undamaged.
- The steering mechanism must operate smoothly without excessive free play or roughness.
- There should be no excessive free play or roughness in the wheel bearings, but there should be sufficient free play to prevent tightness or binding.
- There should be no major fluid leaks from the shock absorbers and, when each corner of the car is depressed, it should rise and then settle in its normal position.
- The driveshafts and propeller shaft (on rear wheel drive cars) must not be damaged or distorted.

Brakes
- It should be possible to operate the handbrake without excessive force or excessive movement of the lever, and it must not be possible for the lever to release unintentionally. The handbrake must be able to lock the relevant wheels on which it operates.
- The brake pedal must not be damaged, and when the pedal is pressed, resistance should be felt near the top of its travel – hard resistance should be felt, and the pedal should not move down towards the floor. The resistance should be firm and not spongy.
- There should be no signs of any fluid leaks anywhere in the braking system.
- The wheels should turn freely when the brakes are not applied.
- When the brakes are applied, the brakes on all four wheels must work, and the car must stop evenly in a straight line without pulling to one side.

VAUXHALL ASTRA & BELMONT

- All the braking system components must be secure, and there must be no signs of excessive wear or corrosion.

Tyres and wheels

- The tyres must be in good condition, and there must be no signs of excessive wear or damage (refer to *'Regular checks'* on page 82).
- Tyres at the same end of the car must be of the same size and type. Always fit radial tyres; crossply tyres are not suitable for modern cars.

Seatbelts

- The mountings must be secure, and must not be loose or excessively corroded.
- The seatbelt webbing should not be frayed or damaged (this must be checked along the full length of the belts).
- When a seatbelt is fastened, the locking mechanism should hold securely and should release when intended.
- In the case of inertia reel seatbelts, the retractors should work properly when the belts are released.

General

- The windscreen wipers and washers must work properly, and the wiper blades must clear the windscreen without smearing.
- The swept area of the windscreen must have a zone 290mm wide, centred on the steering wheel, free of impact damage greater than 10mm across; this zone must be not encroached by any stickers by more than 10mm. Similar restrictions apply to the remainder of the swept area. Here, the damage or obstruction must not exceed 40mm across.
- The horn must work properly, be loud enough to be heard by other road users and emit a constant tone. Two-tone horns, which alternate between the two frequencies (i.e. they warble), are illegal.
- The exhaust mountings must be secure, and the system itself must be free from leaks and serious corrosion.
- There must be no serious corrosion or damage to the vehicle structure. Corrosion of body panels will not necessarily cause a car to fail the test, but there should be no serious corrosion of any of the 'load bearing' structural components.
- Door handles and locks should work properly from both inside and outside; boot lids and tailgates must close securely; seats and spare tyres must be securely mounted; petrol caps must seal and fasten properly.
- Registration plates must be securely fitted, front and rear, clearly legible, not obscured by items such as a tow bar, with the letters and figures correctly formed and spaced. (Letters or figures formed or positioned so that they are likely to be mis-read, i.e. to personalise a number plate, are now illegal – the vehicle will be failed.)
- Vehicles must display a legible Vehicle Identification Number (VIN). This is either a plate secured to the body or chassis in the engine compartment or a number stamped or etched on the vehicle.
- The exhaust gas emissions level must be within certain limits depending on the age of the car (special equipment is required to measure exhaust gas emissions), and the MOT test also includes a visual check for excessive exhaust smoke when the engine is running. Generally, the car should meet the emission regulations if regular servicing has been carried out.

VAUXHALL ASTRA & BELMONT

FAULT FINDING

The following table isn't intended as an exhaustive guide to fault finding, but it summarises some of the more common faults which may crop up during a car's life.

Hopefully, the table should help you to find the cause of a problem, even if you can't cure it yourself.

When confronted with a fault, try to think calmly and logically about the symptom(s), and you should be able to work out what the fault can't be! Check one item at a time, otherwise if you do clear the fault, you may not know what was causing it.

The commonest cause of difficulty is starting, especially in the Winter. Make sure that your battery is kept fully charged, and that the ignition components are in good condition, clean and dry (refer to *Servicing* on page 112).

Further details of fault diagnosis, along with comprehensive renewal procedures for most components can be found in our Owners Workshop Manual for your particular model (OWM 635 for 'Mark 1' models, or OWM 1136 for 'Mark 2' and Belmont models)

SYMPTOM	POSSIBLE CAUSES
ENGINE	
Starter motor doesn't turn, and headlights don't come on	● Flat battery ● Loose, dirty or corroded battery connections ● Other electrical or wiring fault
Starter motor doesn't turn, and headlights dim	● Battery charge low ● Loose, dirty or corroded battery connections ● Faulty starter motor ● Other electrical or wiring fault ● Seized engine
Starter motor doesn't turn, and headlights are bright	● Loose or dirty starter motor connections ● Faulty starter motor or ignition switch ● Gear selector lever not in position **N** or **P** (automatic transmission only) ● Other electrical or wiring fault
Starter motor spins, but doesn't turn engine	● Faulty starter motor ● Engine fault (ie worn or damaged flywheel ring gear)
Engine turns slowly but won't start	● Battery charge low ● Electrical or wiring fault ● Wrong grade of engine oil (refer to *'Service specifications'* on page 73)

FAULT FINDING

SYMPTOM	POSSIBLE CAUSES
ENGINE (continued)	
Engine turns but won't fire, or engine starts but won't keep running	● Fuel tank empty ● Damp or dirty ignition system components ● Dirty or loose ignition system connections ● Incorrectly-adjusted or faulty spark plugs (refer to *'Servicing'* on page 112) ● Fuel system fault ● Other electrical or wiring fault
Engine idles, but stalls when accelerator pedal is depressed	● Blocked or dirty air cleaner filter element (refer to *'Servicing'* on page 119) ● Fuel system fault ● Ignition system fault ● Electrical fault
Poor acceleration, misfiring or lack of power	● Incorrectly-adjusted or faulty spark plugs (refer to *'Servicing'* on page 112) ● Incorrect engine valve clearances ● Fuel system fault ● Ignition system fault ● Electrical fault
Engine continues to run when ignition is switched off	● Engine overheating (possibly due to low coolant level – refer to *'Regular checks'* on page 80) ● Wrong type of spark plugs fitted (refer to *'Servicing'* on page 112) ● Wrong grade of petrol ● Fuel system fault ● Excessive build-up of carbon inside engine ('decoke' required) ● Faulty ignition switch ● Other electrical or wiring fault
Engine doesn't reach normal operating temperature	● Faulty cooling system thermostat ● Faulty temperature gauge or sensor

FAULT FINDING

SYMPTOM	POSSIBLE CAUSES
ENGINE (continued)	
Engine overheats, or temperature gauge reads too high	• Airflow to radiator obstructed • Blocked radiator, hose or engine coolant passage • Coolant or engine oil level low (refer to *'Regular checks'* on page 80) • Coolant hose(s) leaking, worn or damaged • Faulty cooling system thermostat • Faulty water pump • Faulty water pump drive (1.2 litre engine) • Faulty temperature gauge or sensor • Faulty electric cooling fan or switch
Ignition warning light comes on when engine is running	• Alternator drivebelt loose or broken (refer to *'Servicing'* on page 116) • Faulty alternator • Other electrical or wiring fault
Oil pressure warning light comes on when engine is running	• Oil level below 'MIN' mark on dipstick – refer to *'Regular checks'* on page 80 (if oil level is correct, and light is still on, seek advice, but **don't** start the engine) • Oil leak • Faulty oil pressure switch • Badly-worn engine components
GEARBOX, TRANSMISSION AND CLUTCH	
Difficulty in engaging gear (manual gearbox)	• Low gearbox oil level (refer to *'Servicing'* on page 114) • Worn or faulty gearbox components • Worn or faulty clutch mechanism • Engine idle speed too high
Clutch slips (car does not accelerate when engine revs increase – manual gearbox)	• Faulty clutch mechanism • Contaminated or worn clutch
Car gearchange is harsh (automatic transmission)	• Fault in transmission
Car pull away/gear change sluggish (automatic transmission)	• Low transmission fluid level (refer to *'Servicing'* on page 115) • Fault in transmission

FAULT FINDING

SYMPTOM	POSSIBLE CAUSES
BRAKES	
Brakes feel 'spongy'	● Air in brake hydraulic system ● Fluid leak in brake system
Excessive brake pedal travel	● Faulty rear brake mechanism ● Air in brake hydraulic system ● Fluid leak in brake system
Brakes require excessive pedal pressure	● Damp, dirty or contaminated brake components ● Fluid leak in brake system ● Faulty or seized brake components ● Faulty brake servo
Car pulls to one side	● Incorrect tyre pressure(s) or uneven tyre wear (refer to 'Regular checks' on page 82) ● Faulty or seized brake components ● Worn or contaminated brake friction material ● Incorrect wheel alignment ● Worn or faulty steering components ● Worn or faulty suspension components
Brakes squeal and/or judder	● Badly-worn or corroded brake components ● Incorrectly-assembled brake components ● Contaminated brake friction material ● Worn or faulty suspension components ● Worn or faulty steering components
Brake fluid level warning light comes on	● Low brake fluid level (refer to 'Regular checks' on page 81) ● Faulty fluid level sensor or wiring

FAULT FINDING

SYMPTOM	POSSIBLE CAUSES
SUSPENSION AND STEERING	
Car becomes heavy to steer	● Low tyre pressure(s) (refer to *'Regular checks'* on page 82) ● Incorrect wheel alignment ● Worn or faulty steering components ● Worn or faulty suspension components
Car pulls to one side	● Refer to *'Brakes'* section of table
Car wanders	● Incorrect tyre pressure(s) or uneven tyre wear (refer to *'Regular checks'* on page 82) ● Car unevenly loaded ● Incorrect wheel alignment ● Worn or faulty suspension components ● Worn or faulty steering components
Car vibrates when driving	● Loose wheel bolt(s) ● Wheel(s) out of balance ● Worn or damaged driveshaft ● Worn or faulty suspension components ● Worn or faulty steering components ● Worn or faulty brake components
Hard or choppy ride	● Incorrect tyre pressure(s) (refer to *'Regular checks'* on page 82) ● Worn or faulty suspension components
Car leans excessively when cornering	● Roof rack overloaded ● Car unevenly loaded ● Worn or faulty suspension components
Uneven tyre wear	● Incorrect tyre pressure(s) (refer to *'Regular checks'* on page 82) ● Incorrect wheel alignment ● Wheel(s) out of balance ● Worn or faulty suspension components ● Worn or faulty steering components ● Brakes 'grabbing'

FAULT FINDING

SYMPTOM	POSSIBLE CAUSES
ELECTRICS	
Electrical systems don't work	• Loose, dirty or corroded battery connections • Faulty earth connection • Battery charge low • Blown fuse (refer to *'Bulb, fuse and relay renewal'* on page 139) • Faulty relay • Fuse link, connecting main wiring loom to battery blown (seek advice)
Direction indicators don't work	• Blown fuse (refer to *'Bulb, fuse and relay renewal'* on page 139) • Faulty earth connection • Faulty direction indicator relay • Faulty direction indicator switch
Bulbs burn out repeatedly	• Faulty connections at light socket • Faulty alternator regulator
All lights dim when engine speed drops to idle	• Loose alternator drivebelt (refer to *'Servicing'* on page 116) • Battery charge low • Faulty alternator

CAR JARGON

This Section should help you to understand some of the odd words of phrases used at garages and by 'car enthusiasts' which you may not be familiar with.

A

ABS – Abbreviation for Anti-lock Braking System. Uses sensors at each wheel to sense when the wheels are about to lock, and releases the brakes to prevent locking. This process occurs many times per second, and allows the driver to maintain steering control when braking hard.
Accelerator pump – A device attached to many *carburettors* which provides a spurt of extra fuel to the carburettor fuel/air mixture when the accelerator pedal is suddenly pressed down.
Additives – Compounds which are added to petrol and oil to improve their quality and performance.
Advance and retard – A system for altering the *ignition timing*.
AF – An abbreviation of 'Across Flats', the way in which many nuts, bolts and spanners are identified. AF is usually preceded by an Imperial unit of measurement – eg ½ in AF. Unless otherwise stated, all metric measurements are assumed to be AF, so the abbreviation is not normally used for metric nuts, bolts and spanners.
Air cooling – Alternative method of engine cooling in which no water is used. An engine-driven fan forces air at high speed over the surfaces of the engine.
ALB – Abbreviation for Anti-Lock Braking System. See *'ABS'*.
Alternator – A device for converting rotating mechanical energy into electrical energy. In modern cars, it has superseded the dynamo for charging the battery because of its much greater efficiency.
Ammeter – A device for measuring electrical current – the current supplied to the battery by the alternator, or drawn from the battery by the car's electrical systems.
Antifreeze – A chemical mixed with the water in the cooling system to lower the temperature at which the coolant freezes, and in modern cars to prevent corrosion of the metal in the cooling system.
Anti-roll bar – A metal bar mounted transversely across the car, connecting the two sides of the suspension, which counteracts the natural tendency for the car to lean when cornering.
Aquaplaning – A word used to describe the action of a tyre skating across the surface of water.
Automatic transmission – A type of *gearbox* which selects the correct gear ratio automatically according to engine speed and load.
Axle – Spindle on which a wheel revolves.

B

Balljoint – A ball-and-socket type joint used in steering and *suspension* systems, which allows relative movement in more than one plane.
Battery condition indicator – A device for measuring electrical voltage (a voltmeter) connected via the ignition switch to the car battery. Unlike an *ammeter*, it gives an indication that a battery is close to failing. **Also**, most 'maintenance-free' batteries have a battery condition indicator fitted to their casing, which consists of a small disc which changes its colour when the battery is close to failure and requires renewal.
Bearing – Metal or other hard wearing surface against which another part moves, and which is designed to reduce friction and wear (bearings are usually lubricated).

CAR JARGON

Bendix drive – A device on some types of starter motor which allows the motor to drive the engine for starting, then disengages when the engine starts to run.
BHP – see *Horsepower*.
Big end – The end of a *connecting rod* which is attached to the *crankshaft*. It incorporates a *bearing* and transmits the linear movement of the connecting rod to the crankshaft.
Bleed nipple (or valve) – A hollow screw which allows air or fluid to be bled out of a system when it is loosened.
Brake caliper – The part of a *disc brake* system which houses the *brake pads* and the hydraulically-operated pistons.
Brake disc – A rotating disc, coupled to a roadwheel, which is clamped between hydraulically operated friction pads in a *disc brake* system.
Brake fade – A temporary loss of braking efficiency due to overheating of the brake friction material.
Brake pad – The part of a *disc brake* system which consists of the friction material and a metal backing plate.
Brake shoe – The part of a *drum brake* system which consists of the friction material and a curved metal former.
Breather – A device which allows fresh air into a system or allows contaminated air out.
Bucket tappet – A bucket shaped component used in some engines to transfer the rotary movement of the *camshaft* to the up-and-down movement required for *valve* operation.
Bump stop – A hard piece of rubber used in many *suspension* systems to prevent the moving parts from contacting the body during violent suspension movements.

C

Camber angle – The angle at which the front wheels are set from the vertical, when viewed from the front of the car. Positive camber is the amount in degrees which the wheels are tilted out at the top.
Cam follower – A piece of metal used to transfer the rotary movement of the *camshaft* to the up-and-down movement required for *valve* operation.
Camshaft – A rotating shaft driven from the *crankshaft* with lobes or cams used to operate the engine *valves* via the *valve gear*.
Carbon leads – Ignition HT leads incorporating carbon (black fibres) which eliminates the need for separate radio and TV *suppressors*.
Carburettor – A device which is used to mix air and fuel in the proportions required for burning by the engine under all conditions of engine running.
Castor angle – The angle between the front wheels pivot points and a vertical line when viewed from the side of the car. Positive castor is when the axis is inclined rearwards.
Catalytic converter – A device incorporated in the exhaust system which speeds up the natural decomposition of the exhaust gases, and reduces the amount of harmful gases released into the atmosphere. Cars fitted with catalytic converters must be operated on *unleaded petrol*, as *leaded petrol* will destroy the catalyst.
Centrifugal advance – System of ignition *advance and retard* incorporated in many *distributors* in which weights rotating on a shaft alter the ignition timing according to engine speed.
Choke – This has two common meanings. It is used to describe the device which shuts off some of the air in a *carburettor* during cold starting (in order to provide extra fuel), and it may be manually or automatically operated. It's also used as a general term to describe a carburettor throttle bore.
Clutch – A friction device which allows two rotating components to be coupled together smoothly, without the need for either rotating component to stop moving.
Coil spring – A spiral coil of spring steel used in many *suspension* systems.
Combustion chamber – Shaped area in the *cylinder head* into which the fuel/air mixture is compressed by the *piston* and where the spark from the *spark plug* ignites the mixture.
Compression ratio (CR) – A term used to describe the amount by which the fuel/air mixture is compressed as a *piston* moves from the bottom to the top of its travel, and expressed as a number. For example an 8.5:1 compression ratio means that the volume of fuel/air mixture above the piston when the piston is at the bottom of its stroke is 8.5 times

CAR JARGON

that when the piston is at the top of its stroke.
Compression tester – A special type of pressure gauge which can be screwed into a *spark plug* hole, which measures the pressure in the cylinder when the engine is turning but not firing. This gives an indication of engine wear or possible leaks.
Condenser (capacitor) – A device in a *contact breaker point distributor* which stores electrical energy and prevents excessive sparking at the contact breaker points.
Connecting rod ('con-rod') – Metal rod in the engine connecting a *piston* to the *crankshaft*.
Constant velocity (CV) joint – A joint used in *driveshafts*, where the instantaneous speed of the input shaft is exactly the same as the instantaneous speed of the output shaft at any angle of rotation. This does not occur in ordinary *universal joints*.
Contact breaker points – A device in the *distributor* which consists of two electrical points (or contacts) and a cam which opens and closes them to operate the *HT* electrical circuit which provides the spark at the *spark plugs*.
Crankcase – The area of the *cylinder block* below the *pistons* which houses the *crankshaft*.
Crankshaft – A cranked shaft which is driven by the *pistons* and provides the engine output to the *transmission*.
Crossflow cylinder head – A *cylinder head* in which the inlet and exhaust *valves* and *manifolds* are on opposite sides.
Crossply tyre – A tyre whose construction is such that the weave of the fabric material layers is running diagonally in alternately opposite directions to a line around the circumference of the tyre.
Cubic capacity – The total volume within the *cylinders* of an engine which is swept by the movement of the *pistons*.
CVH – A term applied by the Ford Motor Company to their overhead camshaft engines which incorporate a hemispherical *combustion chamber*. CVH means Compound Valve angle, Hemispherical combustion chamber.
Cylinder – Close fitting metal tube in which a *piston* slides. In the case of an engine, the cylinders may be bored directly into the *cylinder block*, or on some engines, cylinder liners are used which rest in the cylinder block and can be replaced when worn with matching pistons to avoid the requirement for *reboring* the cylinder block.
Cylinder block – The main engine casting which contains the *cylinders*, *crankshaft* and *pistons*.
Cylinder head – The casting at the top of the engine which contains the *valves* and associated operating components.

D

Damper – See *shock absorber*.
Dashpot – An oil-filled *cylinder* and *piston* used as a damping device in SU and Zenith/Stromberg CD type carburettors.
Dead axle (beam axle) – The simplest form of axle, consisting of a horizontal member attached to the car underbody by springs. This arrangement is used for the rear axle on some front-wheel-drive cars.
Decarbonising ('decoking') – Removal of all carbon deposits from the *combustion chambers* and the tops of the *pistons* and *cylinders* in an engine.
De Dion axle – A rear axle consisting of a cranked tube attached to the wheel hubs, with a separately mounted *differential* gear and *driveshafts*. *Suspension* is normally through *coil springs* between the wheel hubs and car underbody.
Derv – Abbreviation for Diesel-Engined Road Vehicle. A term often used to refer to Diesel fuel.
Diaphragm – A flexible membrane used in some components such as fuel pumps. The diaphragm spring used on *clutches* is similar, but is made from spring steel.
Diesel engine – An engine which relies on the heat generated when compressing air to ignite the fuel, and which therefore doesn't need an *ignition system*. Diesel engines have much higher *compression ratios* than petrol engines, normally around 20:1.
Differential – A system of gears (generally known as a crownwheel and pinion) which allows the *torque* provided by the engine to be applied to both driving wheels. The differential divides the torque proportionally between the driving wheels to allow one wheel to turn faster than the other, for example during cornering.

CAR JARGON

DIN – This stands for Deutsche Industrie Norm (German Industry Standard), which provides international standards for measuring engine power, torque, etc.
Disc brake – A brake assembly where a rotating disc is clamped between hydraulically operated friction pads.
Distributor – A device used to distribute the HT current to the individual *spark plugs*. The distributor may also contain the *advance and retard* mechanism. On some older cars, the distributor also contains the *contact breaker points* assembly.
Distributor cap – Plastic cap which fits on top of the *distributor* and contains electrodes, in which the *rotor arm* rotates to distribute the HT spark voltage to the correct *spark plug*.
DOHC – Abbreviation for Double Overhead Camshaft (see *'Twin-cam'*).
Doughnut – A term used to describe the flexible rubber coupling used in some *driveshafts*.
Driveshaft – Term usually used to describe the shaft (normally incorporating *universal* or *constant velocity joints*), which transmits drive from a *differential* to one wheel. More commonly found in front-wheel-drive cars.
Drive train – A collective term used to describe the *clutch/gearbox/transmission* and the other components used to transmit drive to the wheels.
Drum brake – A brake assembly with friction linings on 'shoes' running inside a cylindrical drum attached to the wheel.
Dual circuit brakes – A *hydraulic* braking system consisting of two separate fluid circuits, so that if one circuit becomes inoperative, braking power is still available from the other circuit.
Dwell angle – A measurement which corresponds to the number of degrees of *distributor* shaft rotation during which the *contact breaker points* are closed during the ignition cycle of one *cylinder*. The angle is altered by adjusting the contact breaker points gap.

E

Earth strap – A flexible electrical connection between the battery and a car earth point, or between the engine/*gearbox* and the car body to provide a return current path flow to the battery.
EFI – Abbreviation for Electronic *Fuel Injection*.
Electrode – An electrical terminal, eg in a *spark plug* or *distributor cap*.
Electrolyte – A current-conducting solution inside the battery (consisting of water and sulphuric acid in the case of a car battery).
Electronic ignition – An *ignition system* incorporating electronic components in place of *contact breaker points*, which can produce a much higher spark voltage than a contact breaker system, and is less affected by worn components.
Emission control – The reduction or prevention of the release into the atmosphere of poisonous fumes and gases from the engine and fuel system of a car. Required to different degrees by the laws of some countries, and achieved by engine design and the use of special devices and systems.
Epicyclic gears (planetary gears) – A gear system used in many *automatic transmissions* where there is a central 'sun' gear around which smaller 'planet' gears rotate.
Exhaust gas analyser – An instrument used to measure the amount of pollutants (mainly carbon monoxide) in a car's exhaust gases.
Expansion tank – A container used in many cooling systems to collect the overflow from the car's cooling system as the coolant heats up and expands.

F

Filter – A device for removing foreign particles from air, fuel or oil.
Final drive – A collective term (often expressed as a gear ratio) for the crownwheel and pinion (see *Differential*).
Flat engine – Form of engine design in which the *cylinders* are opposed horizontally, usually with an equal number on each side of the central *crankshaft*.
Float chamber – The part of a *carburettor* which contains a float and *needle valve* for

CAR JARGON

controlling the fuel level in the reservoir.
Flywheel – A heavy rotating metal disc attached to the *crankshaft* and used to smooth out the pulsing from the *pistons*.
Four stroke (cycle) – A term used to describe the four operating strokes of a *piston* in a conventional car engine. These are (1) Induction – drawing the air/fuel mixture into the engine as the piston moves down; (2) Compression – of the fuel/air mixture as the piston rises; (3) Power stroke – where the piston is forced down after the fuel/air mixture has been ignited by the *spark plug*, and (4) Exhaust stroke – where the piston rises and pushes the burnt gases out of the *cylinder*. During these operations, the inlet and exhaust *valves* are opened and closed at the correct moment to allow the fuel/air mixture in, the exhaust gases out, or to provide a gas-tight compression chamber.
Fuel injection – A method of injecting fuel into an engine. Used in *Diesel engines* and also on some petrol engines in place of a *carburettor*.
Fuel injector – Device used on *fuel injection* engines to inject fuel directly or indirectly into the *combustion chamber*. Some fuel injection systems use a single fuel injector, while some systems use one fuel injector for each *cylinder* of the engine.

G

Gasket – Compressible material used between two surfaces to provide a leakproof joint.
Gearbox – A group of gears and shafts installed in a housing, positioned between the *clutch* and the *differential*, and used to keep the engine within its safe operating speed range as the speed of the car changes.

H

Half-shaft – A *driveshaft* used to transmit the drive from the *differential* to one of the rear wheels.
Hardy-Spicer joint (Hooke's or Cardan joint) – See *Universal joint*.
Helical gears – Gears in which the teeth are cut at an angle across the circumference of the gear to give a smoother mesh between gears and quieter running.
Horsepower – A measurement of power. Brake Horsepower (BHP) is a measure of the power required to stop a moving body.
HT – Abbreviation of High Tension (meaning high voltage) used to describe the *spark plug* voltage in an *ignition system*.
Hub carrier – A component usually found at each front corner of a car which carries the wheel and brake assembly, and to which the *suspension* and steering components are attached.
Hydraulic – A term used to describe the operation of a system by means of fluid pressure.
Hypoid gear – A gear with curved teeth which transmits drive through a right-angle, where the centreline of the drive gear is offset from the centreline of the driven gear. The meshing action of hypoid gears allows a larger and therefore stronger drive gear, and the meshing noise is reduced in comparison with conventional gears.

I

Independent suspension – A *suspension* system where movement of one wheel has no effect on the movement of the other, eg independent front suspension.
Ignition coil – An electrical coil which forms part of the *ignition system* and which generates the *HT* voltage.
Ignition system – The electrical system which provides the spark to ignite the air/fuel mixture in the engine. Normally the system consists of the battery, *ignition coil*, *distributor*, ignition switch, *spark plugs* and wiring.
Ignition timing – The time in the *cylinder* firing cycle at which the ignition spark (provided by the *spark plug*) occurs. The spark timing is normally a few degrees of *crankshaft* rotation before the *piston* reaches the top of its stroke, and is expressed as a number of degrees before top-dead-centre (BTDC).
Inertia reel – Automatic type of seat belt mechanism which allows the wearer to move freely in normal use, but which locks on sensing either sudden deceleration or a sudden movement of the wearer.

CAR JARGON

In-line engine – An engine in which the *cylinders* are positioned in one row as opposed to being in a *flat* or *vee* configuration.

J

Jet – A calibrated nozzle or orifice in a *carburettor* through which fuel is drawn for mixing with air.

Jump leads – Heavy electric cables fitted with clips to enable a car's battery to be connected to another battery for emergency starting.

K

Kerb weight – The weight of a car, unladen but ready to be driven, ie with enough fuel, oil, etc, to travel an arbitrary distance.

Kickdown – A device used on *automatic transmissions* which allows a lower gear to be selected for improved acceleration by fully depressing the accelerator.

Kingpin – A device which allows the front wheel of a car to swivel about a near vertical axis.

Knocking – See 'Pinking'.

L

Laminated windscreen – A windscreen which has a thin plastic layer sandwiched between two layers of toughened glass. It will not shatter or craze when hit.

Lead-free petrol – Contains no lead. It has no lead added during manufacture, and the natural lead content is refined out. This type of petrol is not currently available for general use in the UK, and should not be confused with *unleaded* petrol.

Leaded petrol – Normal 4-star petrol. Has a low amount of lead added during manufacture, in addition to the natural lead found in crude oil.

Leading shoe – A *drum brake* shoe of which the leading end (the one moved by the operating *pistons*) is reached first by a given point on the drum during normal forward rotation. A simple drum brake will have one leading and one trailing (the opposite) shoe.

Leaf spring – A spring commonly used on cars with a *live axle*, consisting of several long curved steel plates clamped together.

Limited slip differential – A type of *differential* which prevents one wheel from standing still while the other wheel spins excessively. Often used on high-performance cars.

Live axle – An axle through which power is transmitted to the rear wheels.

Loom – A complete car wiring system or section of a wiring system consisting of all the wires of correct length, etc, to wire up the various circuits.

LT – Abbreviation of Low Tension (meaning low voltage), used to describe battery voltage in the *ignition system*.

M

MacPherson strut – An independent front *suspension* system where the swivelling, springing and shock absorbing action of the wheels is dealt with by a single assembly.

Manifold – A device used for ducting the air/fuel mixture to the engine (inlet manifold), or the exhaust gases from the engine (exhaust manifold).

Master cylinder – A *cylinder* containing a *piston* and *hydraulic* fluid, directly coupled to a foot pedal (eg brake or *clutch* master cylinder). Used for transmitting pressure to the brake or clutch operating mechanism.

Metallic paint – Paint finish incorporating minute particles of metal to give added lustre to the colour.

Multigrade – Lubricating oil whose *viscosity* covers that of several single grade oils, making it suitable for use over a wider range of operating conditions.

N

Needle bearing – Type of *bearing* in which needle or cone-shaped rollers are used around the circumference to reduce friction.

Needle valve – A component of the *carburettor* which restricts the flow of fuel or fuel/air mixture according to the position of the valve in an orifice or *jet*.

CAR JARGON

Negative earth – Electrical system (almost universally adopted) in which the negative terminal of the car battery is connected to the car body. The polarity of all the electrical equipment is determined by this.

O

Octane rating – A scale rating for grading petrol.
ohc (overhead cam) – Describes an engine in which the *camshaft* is situated above the *cylinder head*, and operates the *valve gear* directly.
ohv (overhead valve) – Describes an engine which has its *valves* in the *cylinder head*, but with the *valve gear* operated by *pushrods* from a *camshaft* situated lower in the engine.
Oil cooler – Small *radiator* fitted in the oil circuit and positioned in a cooling airflow to cool the oil. Used mainly on high-performance engines.
Overdrive – A device coupled to a car's *gearbox* which raises the output gear ratio above the normal 1:1 of top gear.
Oversteer – A tendency for a car to turn more tightly into a corner than intended.

P

PCV (Positive Crankcase Ventilation) – A system which allows fumes and vapours which build up in the *crankcase* to be drawn into the engine for burning.
Pinion – A gear with a small number of teeth which meshes with one having a larger number of teeth.
Pinking – A metallic noise from the engine often caused by the *ignition timing* being too far *advanced*. The noise is the result of pressure waves which cause the *cylinder* walls to vibrate when the ignited fuel/air mixture is compressed.
Piston – Cylindrical component which slides in a closely-fitting metal tube or *cylinder* and transmits pressure. The pistons in an engine, for example, compress the fuel/air mixture, transmit the power to the *crankshaft*, and push the burnt gases out through the exhaust *valves*.

Piston ring – Hardened metal ring which is a spring fit in a groove running round the *piston* to ensure a gas-tight seal between the piston and *cylinder* wall.
Positive earth – The opposite of *negative earth*.
Power steering – A steering system which uses *hydraulic* fluid pressure (provided by an engine-driven pump) to reduce the effort required to steer the car.
Pre-ignition – See 'Pinking'.
Propeller shaft – The shaft which transmits the drive from the *gearbox* to the rear axle in a front-engined rear-wheel-drive car.
Pushrod – A rod which is moved up and down by the rotary motion of the *camshaft* and operates the *rocker arms* in an *ohv* engine.

Q

Quarter light – A triangular window mounted in front or behind the main front or rear windows, usually in the front door, or behind the rear door.
Quartz-halogen bulb – A bulb with a quartz envelope (instead of glass), filled with a halogen gas. Gives a brighter, more even spread of light than an ordinary bulb.

R

Rack and pinion – Simplest form of steering mechanism which uses a *pinion* gear to move a toothed rack.
Radial ply tyre – A tyre in which the fabric material plies are arranged laterally, at right angles to the circumference.
Radiator – Cooling device through which the engine coolant is passed, situated in an air flow and consisting of a system of fine tubes and fins for rapid heat dissipation.
Radius arms (rods) – Locating arms sometimes used with a *live axle* to positively locate it in the fore-and-aft direction.
Rebore – The process of enlarging the *cylinder* bores to a very accurately specified measurement in order to fit new *pistons* to overcome wear in the engine. Not normally necessary unless the engine has covered a very high mileage.

CAR JARGON

Recirculating ball steering – A derivative of *worm and nut* steering, where the steering shaft motion is transmitted to the steering linkage by balls running in the groove of a worm gear.
Rev counter – See *Tachometer*.
Rocker arm – A lever which rocks on a central pivot, with one end moved up and down by the *camshaft*, and the other end operating an inlet or exhaust *valve*.
Rotary engine – See '*Wankel engine*'.
Rotor arm – A rotating arm in the *distributor* which distributes the *HT* spark voltage to the correct *spark plug*.
Running on – A tendency for an engine to keep on running after the ignition has been switched off. Often caused by a badly maintained engine or the use of an incorrect grade of fuel.

S

SAE – Society of Automotive Engineers (of America). Lays down international standards for the classification of engine performance and many other specifications, but is most commonly used to classify oils.
Safety rim – A special wheel rim shape which prevents a deflated tyre from rolling off the wheel.
Sealed beam – A sealed headlamp unit where the filament is an integral part and cannot be renewed separately.
Semi-trailing arm – A common form of independent rear *suspension*.
Servo – A device for increasing the normal effort applied to a control.
Shock absorber – A device for damping out the up-and-down movement of the *suspension* when the car hits a bump in the road.
Spark plug – A device with two *electrodes* insulated from each other by a ceramic material, which screws into an engine *combustion chamber*. When the *HT* voltage is applied to the plug terminal, a spark jumps across the electrodes and ignites the fuel/air mixture.
Squab – Another name for a seat cushion.
Steel-braced tyre – Tyre in which extra plies containing steel cords are incorporated with the fabric plies to give added strength.

Steering arm (knuckle) – A short arm on the front *hub carrier* to which the steering linkage connects.
Steering gear – A general term used to describe the steering components, usually refers to a steering rack-and-pinion assembly.
Steering rack – See *Rack and pinion*.
Stroboscopic light – A light switched on and off by the engine *ignition system* which is used for checking the *ignition timing* when the engine is running.
Stroke – The total distance travelled by a single *piston* in its *cylinder*.
Stub axle – A short axle which carries one wheel.
Subframe – A small frame which is mounted on the car's body, and carries the *suspension* and/or the *drivetrain* assemblies.
Sump – The main reservoir for the engine oil.
Supercharger – A device which uses an engine-driven turbine (usually driven by a belt or gears from the *crankshaft*) to drive a compressor which forces air into the engine, providing increased fuel/air mixture flow, and therefore increased engine efficiency. Sometimes used on high-performance engines.
Suppressor – A device which is used to reduce or eliminate electrical interference caused by the *ignition system* or other electrical components.
Suspension – A general term used to describe the components which suspend the car body on its wheels.
Swing axle – A *suspension* arm which is pivoted near the front-to-rear centreline of the car, and which allows the wheel to swing vertically about that pivot point.
Synchromesh – A device in a *gearbox* which synchronises the speed of one gear shaft with another to produce smooth, noiseless engagement of the gears.

T

Tachometer – Also known as a rev counter, indicates engine speed in revolutions per minute (rpm).
Tappet – A term often used to refer to the component which transmits the rotary *camshaft* movement to the up-and-down movement required for *valve* operation.

CAR JARGON

Thermostat – A device which is sensitive to changes in engine coolant temperature, and opens up an additional path for coolant to flow through the *radiator* (to increase the cooling) when the engine has warmed up.
Tie-rod – A rod which connects the *steering arms* to the *steering gear*.
Timing belt – Fabric or rubber belt engaging on sprocket wheels and driving the *camshaft* from the *crankshaft*.
Timing chain – Metal flexible link chain engaging on sprocket wheels and driving the *camshaft* from the *crankshaft*.
Timing marks – Marks normally found on the *crankshaft* pulley or the *flywheel* and used for setting the ignition firing point with respect to a particular *piston*.
Toe-in/toe-out – The amount by which the front wheels point inwards or outwards from the straight-ahead position when steering straight ahead.
Top Dead Centre (TDC) – The point at which a *piston* is at the top of its *stroke*.
Torque – The turning force generated by a rotating component.
Torque converter – A coupling where the driving *torque* is transmitted through oil. At low speeds there is very little transfer of torque from the input to the output. As the speed of the input shaft increases, the direction of fluid flow within a system of vanes changes, and torque from the input impeller is transferred to the output turbine. The higher the input speed, the closer the output speed approaches it, until they are virtually the same.
Torsion bar – A metal bar which twists about its own axis, and is used in some *suspension* systems.
Toughened windscreen – A windscreen which when hit, will shatter in a particular way to produce blunt-edged fragments or will craze over but remain intact. A zone toughened windscreen has a zone in front of the driver which crazes into larger parts to reduce the loss of visibility which occurs when toughened windscreens break, but is otherwise similar.
Track rod – See *Tie-rod*.
Trailing arm – A form of independent *suspension* where the wheel is attached to a swinging arm, and is mounted to the rear of the arm pivot.

Transaxle – A combined *gearbox*/axle assembly from which two *driveshafts* transmit the drive to the wheels.
Transmission – A general term used to describe some or all of the *drivetrain* components excluding the engine, most commonly used to describe automatic gearboxes.
Turbocharger – A device which uses a turbine driven by the engine exhaust gases to drive a compressor which forces air into the engine, providing increased fuel/air mixture flow, and therefore increased engine efficiency. Commonly used on high-performance engines.
Twin-cam – Abbreviation for twin overhead *camshafts* (see *'ohc'*). Used on engines with a *crossflow cylinder head*, usually with one camshaft operating the inlet *valves* and the other operating the exhaust valves. Gives improved engine efficiency due to improved fuel/air mixture and exhaust gas flow in the *combustion chambers*.
Two stroke (cycle) – A common term used to describe the operation of an engine where each downward *piston* stroke is a power stroke. The fuel/air mixture is directed to the crankcase where it's compressed by the descending piston and pumped into the *combustion chamber*. As the piston rises, the mixture is compressed and ignited, which forces the piston down. The burnt gases flow from the exhaust port, but the piston is now compressing another fuel/air mixture charge in the crankcase which repeats the cycle.

U

Understeer – A tendency for a car to go straight on when turned into a corner.
Universal joint – A joint that can swivel in any direction whilst at the same time transmitting *torque*. This type of joint is commonly used in *propeller shafts* and some *driveshafts*, but is not suitable for some applications because the input and output shaft speeds are not the same at all positions of angular rotation. The type in common use is known as a Hardy-Spicer, Hooke's or Cardan joint.
Unleaded petrol – Has no lead added during manufacture, but still has the natural lead content of crude oil. Generally available in the

CAR JARGON

UK, most modern cars can use this type of petrol, but seek advice first, as engine adjustments may be required. Engine damage can occur if unleaded petrol is used incorrectly. Not to be confused with *lead-free* petrol which is not currently available in the UK.

Unsprung weight – The part of the car which is not supported by the springs.

V

Vacuum advance – System of ignition *advance and retard* used in some *distributors* where the vacuum in the engine inlet *manifold* is used to act on a *diaphragm* which alters the *ignition timing* as the vacuum changes due to the throttle position.

Valve – A device which opens or closes to allow or stop gas or fluid flow.

Valve gear – A general term used for the components which are acted on by the *camshaft* in order to operate the *valves*.

16-valve – Term used to describe a four *cylinder* engine with four *valves* per cylinder (usually two inlet valves and two exhaust valves). Gives improved engine efficiency due to improved fuel/air mixture and exhaust gas flow in the *combustion chambers*.

Vee engine – An engine design in which the *cylinders* are set in two banks forming a 'V' when viewed from one end. A V8 for example consists of two rows of four cylinders each.

Venturi – A streamlined restriction in the *carburettor* throttle bore which causes a low pressure to occur; this sucks fuel into the air stream to form a vapour suitable for combustion.

Viscosity – A term used to describe the resistance of a fluid to flow. When associated with lubricating oil, it's given an *SAE* number, 10 being a very light oil and 140 being a very heavy oil.

Voltage regulator – A device which regulates the *alternator* output to a predetermined level. On most alternators the voltage regulator is an integral part of the alternator, and regulates the charging current as well as the voltage.

W

Wankel engine – A rotary engine which has a triangular shaped rotor which performs the function of the *pistons* in a conventional engine, and rotates in a housing shaped approximately like a broad-waisted figure of eight. Very few cars use this type of engine.

Wheel balancing – Adding small weights to the rim of a wheel so that there are no out-of-balance forces when the wheel rotates.

Wishbone – An 'A'-shaped *suspension* component, pivoted at the base of the 'A' and carrying a wheel at the apex. Normally mounted close to the horizontal.

Worm and nut steering – A steering system where the lower end of the steering column has a coarse screw thread on which a nut runs. The nut is attached to a spindle which carries the drop arm which, in turn, moves the steering linkage.

LOCAL RADIO FREQUENCIES

A comprehensive network of local radio stations now exists throughout the UK.

Most of these radio stations provide regularly updated reports on traffic flow and road conditions, which can be of great help to drivers. Listening to traffic reports will help you to avoid the inevitable 'jams' which occur during every day driving.

Some car radio/cassette players are now equipped with a 'Radio Data System' (RDS) which will automatically tune into special traffic information signals, interrupting normal radio or tape listening when bulletins are broadcast. An RDS-equipped radio may prove to be a worthwhile investment if you travel by car regularly, particularly when driving on business trips.

The following table provides details of all the local radio frequencies throughout the UK, region-by-region. Where stations broadcast on FM/VHF and AM/MW, it's suggested that the FM/VHF frequency is used wherever possible, as this will usually give better reception.

160 LOCAL RADIO FREQUENCIES

AREA	FM/VHF	AM/MW
AVON		
Bath		
BBC Bristol	104.6	1548
GWR FM	103.0	–
Bristol		
BBC Bristol	94.9	1548
Brunel Classic Gold	–	1260
Galaxy Radio	97.2	–
GWR FM	96.3	–
BEDFORDSHIRE		
Bedford		
BBC Bedfordshire	95.5	1161
Chiltern Radio	96.9	–
SuperGold	–	792
Luton		
BBC Bedfordshire	103.8	630
Chiltern Radio	97.6	–
SuperGold	–	828
BERKSHIRE		
Reading		
210 Classic Gold Radio	–	1431
210 FM	97.0	–
BBC Berkshire	104.4	–
Wokingham		
BBC Berkshire	104.1	–
BIRMINGHAM		
Birmingham		
BBC WM	95.6	1458
BRMB FM	96.4	–
Buzz FM	102.4	–
Xtra AM	–	1152
BORDERS		
Eyemouth		
Radio Borders	103.4	–
Peebles		
Radio Borders	103.1	–
Selkirk		
Radio Borders	96.8	–
BUCKINGHAMSHIRE		
Milton Keynes		
BBC Bedfordshire	104.5	630
Horizon Radio	103.3	–

AREA	FM/VHF	AM/MW
CAMBRIDGESHIRE		
Cambridge		
BBC Cambridge	96.0	1026
CN FM	103.0	–
Peterborough		
BBC Cambridge	95.7	1447
Hereward Radio	102.7	1332
CHANNEL ISLANDS		
Guernsey		
BBC Guernsey	93.2	1116
Jersey		
BBC Jersey	88.8	1026
CHESHIRE		
Echo 96	96.4	–
Chester		
BBC Merseyside	95.8	1485
Congleton		
Signal Cheshire	104.9	–
Macclesfield		
BBC Stoke	94.6	1503
Warrington		
BBC Merseyside	95.8	1485
CLEVELAND		
Middlesbrough		
BBC Cleveland	95.0	1548
TFM	96.6	–
GNR	–	1170
CLWYD		
Prestatyn		
BBC Cymru/Wales	94.2	882
Rhyl		
BBC Cymru/Wales	94.2	882
Wrexham		
BBC Cymru/Wales	93.3	657
MFM 1034	97.1/103.4	–
Marcher Gold	–	1260
CORNWALL		
Isles of Scilly		
BBC Cornwall	96.0	–
Liskeard		
BBC Cornwall	95.2	657
Redruth		
BBC Cornwall	103.9	630
CUMBRIA		
Barrow-in-Furness		
BBC Furness	96.1	837
Kendal		
BBC Cumbria	95.2	–
Windermere		
BBC Cumbria	104.2	–
Workington		
BBC Cumbria	95.6	1458

AREA	FM/VHF	AM/MW
DERBYSHIRE		
Chesterfield		
BBC Sheffield	94.7	1035
Derby		
BBC Derby	94.2	1116
Trent FM	102.8	–
Gem AM	–	945
Matlock		
BBC Derby	95.3	–
DEVON		
Barnstaple		
BBC Devon	94.8	801
Exeter		
BBC Devon	95.8	990
DevonAir	97.0	666/954
South West	103.0	–
Okehampton		
BBC Devon	96.0	801
Plymouth Area		
BBC Devon	103.4	855
Plymouth Sound	97.0	1152
Tavistock		
Radio in Tavistock	96.6	–
Torbay		
BBC Devon	103.4	1458
DevonAir	96.4	666/954
DORSET		
Poole		
BBC Solent	96.1	1359
Bournemouth		
BBC Solent	96.1	1359
2CR Classic Gold	–	828
2CR FM	102.3	–
Weymouth		
BBC Solent	96.1	–
DUMFRIES & GALLOWAY		
Dumfries		
BBC Scotland/Solway	94.7	585
SW Sound	97.2	–
Stranraer		
BBC Scotland	94.7	–
DURHAM		
Darlington		
BBC Newcastle	95.4	–
Durham		
BBC Newcastle	95.4	1458
DYFED		
Aberystwyth		
BBC Cymru/Wales	93.1	882

VAUXHALL ASTRA & BELMONT

LOCAL RADIO FREQUENCIES   161

AREA	FM/VHF	AM/MW
EAST SUSSEX		
Brighton		
BBC Sussex	95.3	1485
South Coast Radio	–	1323
Southern Sound Classic Hits	103.5	–
Eastbourne		
BBC Sussex	104.5	1161
Southern Sound Classic Hits	102.4	–
Hastings		
Southern Sound Classic Hits	97.5	–
Newhaven		
Southern Sound Classic Hits	96.9	–
ESSEX		
Basildon		
BBC Essex	95.3	765
Chelmsford		
BBC Essex	103.5	765
Breeze AM		1431/1359
Essex Radio	102.6	–
Colchester		
BBC Essex	103.5	729
Mellow 1557	–	1557
Harlow		
BBC Essex	–	765
Manningtree		
BBC Suffolk	103.9	–
Southend-on-Sea		
BBC Essex	95.3	1530
Breeze AM		1431/1359
Essex Radio	96.3	–
FIFE		
Kirkcaldy		
BBC Scotland	94.3	810
GLOUCESTERSHIRE		
Gloucester		
BBC Gloucester	104.7	603
Severn Sound	102.4	
Three Counties Radio		774
Stroud		
BBC Gloucester	95.0	603
Severn Sound	103.0	–
GRAMPIAN		
Aberdeen		
BBC Scotland/Aberdeen	93.1	990
North Sound	96.9	1035
Stirling		
Central FM	96.7	–

AREA	FM/VHF	AM/MW
GWENT		
Newport		
Red Dragon Radio	97.4	–
Touch AM		1305
GWYNEDD		
Anglesey		
BBC Cymru/Wales	94.2	882
HAMPSHIRE		
Andover		
210 FM	102.9	–
Basingstoke		
210 FM	102.9	–
BBC Berkshire	104.1	–
Isle of Wight		
BBC Solent	96.1	999/1359
Isle of Wight Radio	–	1242
Portsmouth		
BBC Solent	96.1	999
Oceansound	97.5	–
South Coast Radio	–	1170
Southampton		
BBC Solent	96.1	999
Oceansound	97.5	–
Power FM	103.2	–
South Coast Radio	–	1557
Winchester		
Oceansound	96.7	–
Power FM	103.2	–
HEREFORDSHIRE		
Hereford		
BBC Here/Worc	94.7	819
Radio Wyvern	97.6	954
HERTFORDSHIRE		
Watford		
BBC GLR	94.9	1458
St Albans		
BBC Bedfordshire	103.8	630
Stevenage		
BBC Bedfordshire	103.8	630
HIGHLAND		
Inverness		
BBC Scotland	94.0	810
Moray Firth Radio	97.4	1107
Fort William		
BBC Scotland	93.7	–
Oban		
BBC Scotland	93.3	–
Thurso		
BBC Scotland	94.5	–

AREA	FM/VHF	AM/MW
HUMBERSIDE		
Kingston upon Hull Area		
BBC Humberside	95.9	1485
Classic Gold	–	1161
Viking FM	96.9	–
ISLE OF MAN		
Douglas		
Manx Radio	97.2	1368
Snaefell		
Manx Radio	89.0	–
KENT		
Ashford		
Invicta FM	96.1	–
Canterbury		
BBC Kent	104.2	774
Invicta FM	102.8	–
Dover		
BBC Kent	104.2	774
Invicta FM	97.0	–
East Kent		
Coast Classics	–	603
Folkestone		
BBC Kent	104.2	774
Maidstone & Medway		
BBC Kent	96.7	1035
Coast Classics	–	1242
Invicta FM	103.1	–
Royal Tunbridge Wells		
BBC Kent	96.7	1602
Thanet		
Invicta FM	95.9	–
LANCASHIRE		
Burnley Area		
BBC Lancashire	95.5	855
Blackpool		
BBC Lancashire	103.9	855
Red Rose Gold	–	999
Red Rose Rock FM	97.4	–
Lancaster		
BBC Lancashire	104.5	1557
Preston		
BBC Lancashire	103.9	855
Red Rose Gold	–	999
Red Rose Rock FM	97.4	–
LEICESTERSHIRE		
Leicester		
BBC Leicester	104.9	837
Gem AM	–	1260
Sound FM	103.2	–

VAUXHALL ASTRA & BELMONT

LOCAL RADIO FREQUENCIES

AREA	FM/VHF	AM/MW
LINCOLNSHIRE		
Lincoln Area		
BBC Lincoln	94.9	1368
Lincs FM	102.2	–
LONDON		
Brixton		
Choice FM	96.9	–
Ealing		
Sunrise Radio	–	1413
Greater London		
BBC GLR	94.9	1458
Capital FM	95.8	–
Capital Gold	–	1548
Jazz FM	102.2	–
Kiss FM	100.0	–
LBC Newstalk	97.3	–
Melody Radio	104.9	–
Spectrum Int. Radio	–	558/990
Haringey		
London Greek Radio	103.3	–
WNK	103.3	–
Thamesmead		
London Talkback Radio	–	1152
RTM	103.8	–
Southall		
Sunrise Radio	–	1413
LOTHIAN		
Bathgate		
Radio Forth	97.3	–
Dunfermline		
BBC Scotland	94.3	810
Edinburgh		
BBC Scotland	94.3	810
Max AM	–	1548
Radio Forth	97.3/97.6	–
MANCHESTER		
Manchester		
BBC GMR	95.1	1458
KFM	104.9	–
Piccadilly Gold	–	1152
Piccadilly Key 103	103.0	–
Sunset Radio	102.0	–
MERSEYSIDE		
Liverpool		
BBC Mersey	95.8	1485
City Talk	–	1548
Radio City	96.7	–
MID GLAMORGAN		
Aberdare		
BBC Cymru/Wales	93.6	882
Merthyr Tydfil		
BBC Cymru/Wales	96.8	882

AREA	FM/VHF	AM/MW
MIDDLESEX		
Hounslow		
Sunrise Radio	–	1413
NORFOLK		
King's Lynn		
BBC Norfolk	104.4	873
Norwich Area		
BBC Norfolk	95.1	855
Broadland	102.4	1152
NORTHAMPTONSHIRE		
Kettering		
KCBC	–	1530
Northampton		
BBC Northampton	104.2	1107
Northants	96.6	–
SuperGold	–	1557
NORTHUMBERLAND		
Berwick-upon-Tweed		
BBC Newcastle	96.0	–
Radio Borders	97.5	–
NOTTINGHAM		
Nottingham		
BBC Nottingham	103.8	1521
Trent FM	96.2/96.5	–
Gem AM	–	999
ORKNEY		
Kirkwall		
BBC Scotland	93.7	–
OXFORDSHIRE		
Banbury		
Fox FM	97.4	–
Oxford		
BBC Oxford	95.2	1485
Fox FM	102.6	–
POWYS		
Welshpool		
BBC Cymru/Wales	94.0	882
SHETLAND		
Lerwick		
BBC Scotland	92.7	–
SIBC	96.2	–
SHROPSHIRE		
Ludlow		
BBC Shropshire	95.0	1584
RFM	97.6	–
Shrewsbury		
BBC Shropshire	96.0	–
Beacon Radio	97.2/103.1	–
WABC	–	1017
Telford		
BBC Shropshire	96.0	756
Beacon Radio	97.2/103.1	–
WABC	–	1017

AREA	FM/VHF	AM/MW
SOMERSET		
Mendip Area		
Orchard FM	102.6	–
Wells		
BBC Bristol	95.5	1548
SOUTH GLAMORGAN		
Cardiff		
BBC Cymru/Wales	96.8	882
Red Dragon Radio	103.2	–
Touch AM	–	1359
STAFFORDSHIRE		
Stafford		
Echo 96	96.9	–
Stoke on Trent		
BBC Stoke	94.6	1503
Signal Radio	102.6	1170
STRATHCLYDE		
Ayr		
West Sound	96.7	1035
Girvan		
West Sound	97.5	–
Glasgow		
BBC Scotland	94.3	810
Clyde 1	102.5	–
Clyde 2	–	1152
Radio Clyde FM	97.3	–
Greenock		
BBC Scotland	94.3	810
Kilmarnock		
BBC Scotland	93.9	810
SUFFOLK		
Bury St. Edmunds		
Saxon Radio	96.4	1251
Great Barton		
BBC Suffolk	104.6	–
Ipswich		
BBC Suffolk	103.9	–
Radio Orwell	97.1	1170

LOCAL RADIO FREQUENCIES 163

AREA	FM/VHF	AM/MW
SURREY		
Guildford		
BBC Surrey	104.6	–
Delta Radio	97.1	–
First Gold Radio	–	1476
Premier Radio	96.4	–
Reigate		
Radio Mercury	102.7	1521
TAYSIDE		
Dundee		
Radio Tay	102.8	1161
BBC Scotland	92.7	810
Perth		
Radio Tay	96.4	1584
TYNE & WEAR		
Fenham		
BBC Newcastle	104.4	–
Metro FM	103.0	–
Newcastle Area		
BBC Newcastle	95.4	1458
GNR	–	1152
Metro FM	97.1	–
Sunderland		
Wear FM	103.4	–
ULSTER		
Belfast		
BBC Ulster	94.5	1341
Classic Trax BCR	96.7	–
Cool FM	97.4	–
Downtown Radio	102.6	–
Enniskillen		
BBC Ulster	93.8	873
Downtown Radio	96.6	–
Limavady		
Downtown Radio	96.4	–
Londonderry		
BBC Foyle	93.1	792
Downtown Radio	102.4	–
WARWICKSHIRE		
Leamington Spa		
Mercia FM	102.9	–

AREA	FM/VHF	AM/MW
WEST GLAMORGAN		
Swansea		
BBC Cymru/Wales	93.9	882
Swansea Sound	96.4	1170
WEST MIDLANDS		
Coventry		
BBC CWR	94.8	–
Mercia FM	97.0	–
Radio Harmony	102.6	–
Xtra AM	–	1359
Sutton Coldfield		
BBC Derby	104.5	–
Beacon Radio	97.2/103.1	–
Wolverhampton		
BBC WM	95.6	828
Beacon Radio	97.2/103.1	–
WABC	–	990
WEST SUSSEX		
Crawley		
Radio Mercury	102.7	1521
Horsham		
BBC Sussex	95.1	1368
Worthing		
BBC Sussex	95.3	1485
WESTERN ISLES		
Stornoway (Lewis)		
BBC Scotland	94.2	–
WILTSHIRE		
Chippenham		
BBC Wiltshire Sound	104.3	–
Marlborough		
GWR FM	96.5	–
Salisbury		
BBC Wiltshire Sound	103.5	–
Swindon		
BBC Wiltshire Sound	103.6	1368
Brunel Classic Gold	–	1161
GWR FM	97.2	–
West Wilts		
Brunel Classic Gold	–	936
GWR FM	102.2	–
WORCESTERSHIRE		
Worcester		
BBC Here/Worc	104.0	738
Radio Wyvern	102.8	1530

AREA	FM/VHF	AM/MW
YORKSHIRE NORTH		
Northallerton		
BBC York	104.3	–
Scarborough		
BBC York	95.5	1260
Whitby		
BBC Cleveland	95.8	–
York		
BBC York	103.7	666
YORKSHIRE SOUTH		
Barnsley		
Classic Gold	–	1305
Hallam	102.9	–
Doncaster		
BBC Sheffield	104.1	1035
Classic Gold	–	990
Hallam	103.4	–
Rotherham		
BBC Sheffield	88.6	1035
Hallam	96.1	–
Sheffield		
BBC Sheffield	88.6	1035
Classic Gold	–	1548
Hallam	97.4	–
YORKSHIRE WEST		
Bradford		
Classic Gold	–	1278
Pennine	97.5	–
Sunrise FM	103.2	–
Halifax		
Classic Gold	–	1530
Pennine	102.5	–
Huddersfield		
Classic Gold	–	1530
Pennine	102.5	–
Leeds		
Aire FM	96.3	–
BBC Leeds	92.4	774
Magic 828	–	828

VAUXHALL ASTRA & BELMONT

CONVERSION FACTORS

Length (distance)

Inches (in)	x 25.4	= Millimetres (mm)	x 0.0394	= Inches (in)
Feet (ft)	x 0.305	= Metres (m)	x 3.281	= Feet (ft)
Miles	x 1.609	= Kilometres (km)	x 0.621	= Miles

Volume (capacity)

Cubic inches (cu in; in^3)	x 16.387	= Cubic centimetres (cc; cm^3)	x 0.061	= Cubic inches (cu in; in^3)
Pints (pt)	x 0.568	= Litres (l)	x 1.76	= Pints (pt)
Quarts (qt)	x 1.137	= Litres (l)	x 0.88	= Quarts (qt)
Gallons (gal)	x 4.546	= Litres (l)	x 0.22	= Gallons (gal)

Mass (weight)

Ounces (oz)	x 28.35	= Grams (g)	x 0.035	= Ounces (oz)
Pounds (lb)	x 0.454	= Kilograms (kg)	x 2.205	= Pounds (lb)

Force

Pounds-force (lbf; lb)	x 4.448	= Newtons (N)	x 0.225	= Pounds-force (lbf; lb)
Newtons (N)	x 0.1	= Kilograms-force (kgf; kg)	x 9.81	= Newtons (N)

Pressure

Pounds-force per square inch (psi; lbf/in^2; lb/in^2)	x 0.070	= Kilograms-force per square centimetre (kgf/cm^2; kg/cm^2)	x 14.223	= Pounds-force per square inch (psi; lbf/in^2; lb/in^2)
Pounds-force per square inch (psi; lbf/in^2; lb/in^2)	x 0.068	= Atmospheres (atm)	x 14.696	= Pounds-force per square inch (psi; lbf/in^2; lb/in^2)
Pounds-force per square inch (psi; lbf/in^2; lb/in^2)	x 0.069	= Bars	x 14.5	= Pounds-force per square inch (psi; lbf/in^2; lb/in^2)

CONVERSION FACTORS

Torque (moment of force)

Pounds-force feet (lbf ft; lb ft)	x 0.138	= Kilograms-force metres (kgf m; kg m)	x 7.233	= Pounds-force feet (lbf ft; lb ft)
Pounds-force feet (lbf ft; lb ft)	x 1.356	= Newton metres (Nm)	x 0.738	= Pounds-force feet (lbf ft; lb ft)
Newton metres (Nm)	x 0.102	= Kilograms-force metres (kgf m; kg m)	x 9.804	= Newton metres (Nm)

Power

Horsepower (hp)	x 0.745	= Kilowatts (kW)	x 1.3	= Horsepower (hp)

Velocity (speed)

Miles per hour (miles/hr; mph)	x 1.609	= Kilometres per hour (km/h; kph)	x 0.621	= Miles per hour (miles/hr; mph)

Fuel consumption*

Miles per gallon (mpg)	x 0.354	= Kilometres per litre (km/l)	x 2.825	= Miles per gallon (mpg)

Temperature

Degrees Fahrenheit = (°C x 1.8) + 32

Degrees Celsius (Degrees Centigrade; °C) = (°F – 32) x 0.56

* Note: It is common practice to convert from miles per gallon (mpg) to litres/100 kilometres (l/100 km), where mpg x l/100 km = 282

DISTANCE TABLES

	LONDON	Aberdeen	Aberystwyth	Birmingham	Bournemouth	Brighton	Bristol	Cambridge	Cardiff	Carlisle	Chester	Derby	Dover	Edinburgh	Exeter	Fishguard	Fort William	Glasgow	Harwich	Holyhead
LONDON		806	341	179	169	89	183	84	249	484	298	198	117	608	275	422	818	634	114	420
Aberdeen	501		724	658	903	895	789	742	792	349	571	634	924	198	911	795	254	241	848	692
Aberystwyth	212	450		193	328	425	206	354	177	375	153	222	459	526	325	90	708	525	460	182
Birmingham	111	409	120		237	267	132	161	163	309	117	63	296	460	259	283	642	459	267	246
Bournemouth	105	561	204	147		148	124	251	195	554	351	306	278	705	132	375	887	703	283	463
Brighton	55	556	264	166	92		217	180	288	573	386	286	130	697	275	468	906	723	203	509
Bristol	114	490	128	82	77	135		233	71	439	237	195	314	591	121	251	722	589	314	340
Cambridge	52	461	220	100	156	112	145		290	417	269	155	204	544	351	436	750	566	106	391
Cardiff	155	492	110	101	121	179	44	180		443	237	225	385	587	192	180	776	608	359	341
Carlisle	301	217	233	192	344	356	273	259	275		222	304	602	151	571	444	333	150	523	341
Chester	185	355	95	73	218	240	147	167	147	138		114	415	373	357	243	555	372	375	137
Derby	123	394	138	39	190	178	121	96	140	189	71		315	436	315	312	637	454	261	251
Dover	73	574	285	184	173	81	195	127	239	374	258	196		726	391	539	935	752	203	538
Edinburgh	378	123	327	286	438	433	367	338	365	94	232	271	451		723	616	233	72	650	489
Exeter	171	566	202	161	82	171	75	218	119	355	222	196	243	449		372	904	721	389	460
Fishguard	262	494	56	176	233	291	156	271	112	276	151	194	335	383	231		777	594	533	272
Fort William	508	158	440	399	551	563	480	466	482	207	345	396	581	145	562	483		183	856	671
Glasgow	394	150	326	285	437	449	366	352	378	93	231	282	467	45	448	369	114		673	488
Harwich	71	527	286	166	176	126	195	66	223	325	233	162	126	404	242	331	532	418		497
Holyhead	261	430	113	153	288	316	211	243	212	212	85	156	334	304	286	169	417	303	309	
Hull	168	348	229	143	255	223	226	124	244	155	132	98	251	225	301	285	362	250	181	217
Inverness	537	105	482	445	594	590	528	497	524	253	387	430	601	159	603	529	66	176	563	463
Leeds	190	322	173	111	261	251	206	148	212	115	78	74	269	199	268	229	322	210	214	163
Leicester	98	417	151	39	166	153	112	68	139	208	94	28	171	294	187	207	415	301	134	183
Lincoln	136	382	186	85	217	191	163	86	190	180	123	51	213	259	238	245	387	273	152	208
Liverpool	205	334	104	94	235	260	164	184	164	116	17	88	278	210	239	160	323	211	250	94
Manchester	192	332	133	81	228	247	167	154	172	115	38	58	265	209	242	189	322	210	220	123
Newcastle	271	230	270	209	348	326	295	231	310	58	175	164	344	107	370	326	257	143	297	260
Norwich	107	487	267	155	212	162	221	60	237	287	210	139	169	364	277	331	494	380	63	299
Nottingham	123	388	154	50	189	178	132	84	152	185	87	16	196	265	207	210	392	278	150	172
Oxford	56	464	157	62	92	99	66	79	107	256	128	98	129	341	139	207	461	340	134	202
Penzance	282	683	313	268	193	269	186	329	230	466	333	307	354	560	111	342	673	559	353	397
Plymouth	213	614	244	199	124	213	117	260	161	397	264	238	285	491	42	273	604	490	284	328
Preston	216	302	144	105	256	271	185	198	190	85	49	100	289	179	266	200	292	180	264	125
Sheffield	162	358	158	75	227	217	163	124	176	145	78	36	235	235	248	214	352	240	190	163
Southampton	79	530	202	128	32	60	75	136	119	322	194	164	149	416	111	231	520	415	150	283
Stranraer	410	235	342	301	453	465	382	368	384	109	247	298	483	133	464	385	199	85	434	321
Swansea	196	506	76	126	162	220	85	216	41	289	151	165	280	383	160	71	496	382	264	189
Worcester	114	428	98	26	131	168	60	116	75	211	88	65	187	305	135	155	418	304	182	152
York	196	309	197	134	273	251	215	157	242	116	102	89	269	186	290	253	323	211	223	187

MILES

VAUXHALL ASTRA & BELMONT

DISTANCE TABLES

KILOMETRES

Hull	Inverness	Leeds	Leicester	Lincoln	Liverpool	Manchester	Newcastle	Norwich	Nottingham	Oxford	Penzance	Plymouth	Preston	Sheffield	Southampton	Stranraer	Swansea	Worcester	York	
270	864	306	158	219	330	309	436	172	198	90	454	343	348	261	127	660	315	183	315	LONDON
560	169	518	671	615	538	534	370	784	624	747	1099	988	486	576	853	378	814	689	497	Aberdeen
369	776	278	243	304	167	214	435	430	248	253	504	393	232	254	325	550	122	158	317	Aberystwyth
230	716	179	63	137	151	130	336	249	80	100	431	320	169	121	206	484	203	42	216	Birmingham
410	956	420	267	349	378	367	560	341	304	148	311	200	412	365	51	729	261	211	439	Bournemouth
359	950	404	246	307	418	398	525	261	286	159	433	343	436	349	97	748	354	270	404	Brighton
364	850	332	180	262	264	269	475	356	212	106	299	188	298	262	121	615	137	97	346	Bristol
200	800	238	109	138	296	248	372	97	135	127	529	418	319	200	219	592	348	187	253	Cambridge
393	843	341	224	306	264	277	499	381	245	172	370	259	306	283	192	618	66	121	389	Cardiff
249	407	185	335	290	187	185	93	462	298	412	750	639	137	233	518	175	465	340	187	Carlisle
212	623	126	151	198	27	61	282	338	140	206	536	425	79	126	312	398	243	142	164	Chester
158	692	119	45	82	142	93	264	224	26	158	494	383	161	58	264	480	266	105	143	Derby
404	967	433	275	343	447	426	554	272	315	208	570	459	465	378	240	777	451	301	433	Dover
362	256	320	473	417	338	336	172	586	426	549	901	790	288	378	669	214	616	491	299	Edinburgh
484	970	431	301	383	385	389	595	446	333	224	179	68	428	399	179	747	257	217	467	Exeter
459	851	369	333	394	257	304	525	533	338	333	550	439	322	344	372	620	114	249	407	Fishguard
583	106	518	668	623	520	518	414	795	631	742	1083	972	470	566	837	320	798	673	520	Fort William
402	283	338	484	439	340	338	230	612	447	547	900	789	290	386	668	137	615	489	340	Glasgow
291	906	344	216	245	402	354	478	101	241	216	568	457	425	306	241	698	429	293	359	Harwich
349	745	262	295	335	151	198	418	481	277	325	639	528	201	262	455	517	304	245	301	Holyhead
	615	90	145	61	196	151	190	240	148	262	663	552	180	103	412	425	433	272	63	Hull
382		579	729	673	605	604	428	842	682	816	1159	1049	555	634	925	417	872	747	552	Inverness
56	358		158	108	119	64	148	277	116	272	613	502	90	58	378	360	369	220	39	Leeds
90	453	98		82	187	138	301	187	42	119	480	369	206	105	225	510	282	109	174	Leicester
38	418	67	51		192	137	245	169	56	201	562	451	185	72	307	465	340	192	124	Lincoln
122	376	74	116	119		55	253	360	167	251	563	452	50	119	357	362	270	169	158	Liverpool
94	375	40	86	85	34		212	306	119	230	568	457	48	64	325	360	295	164	103	Manchester
118	266	92	187	152	157	132		414	254	412	774	663	203	206	518	262	517	378	127	Newcastle
149	523	174	116	105	224	190	257		198	224	626	515	354	241	299	637	439	283	293	Norwich
92	424	72	26	35	104	74	158	123		156	512	401	167	64	262	473	283	122	134	Nottingham
163	507	169	74	125	156	143	256	139	97		402	291	272	217	106	587	230	95	291	Oxford
412	720	381	298	349	350	353	481	389	318	250		126	597	560	357	925	436	396	645	Penzance
343	652	312	229	280	281	284	412	320	249	181	78		486	447	246	814	325	285	534	Plymouth
112	345	56	128	115	31	30	126	220	104	169	371	302		113	375	312	322	203	126	Preston
64	394	36	65	45	74	40	128	150	40	135	348	278	70		323	490	336	163	90	Sheffield
256	575	235	140	191	222	202	322	186	163	66	222	153	233	201		679	257	193	398	Southampton
264	259	224	317	289	225	224	163	396	294	365	575	506	194	254	422		641	515	362	Stranraer
269	542	229	175	211	168	183	321	273	176	143	271	202	200	209	160	398		161	418	Swansea
169	464	137	68	119	105	102	235	176	76	59	246	177	126	101	120	320	100		257	Worcester
39	343	24	108	77	98	64	79	182	83	181	401	332	78	56	247	225	260	160		York

VAUXHALL ASTRA & BELMONT

INDEX

A

About this handbook 5
Accidents 39
   Dealing with 39
   First aid 40
   Recording details of 42
   Requirements of the law 42
Air cleaner adjustment 126
Air cleaner element, renewal 119
Alcohol and driving 60
Alternator drivebelt 116
Asbestos, precautions concerning 99
Automatic transmission 23
Automatic transmission fluid level 115

B

Bad weather (driving) 56
Battery electrolyte level 84
Battery precautions 99
Battery, starting when flat 52
Battery terminals, cleaning 126
Bodywork, checks and maintenance 129
Bonnet release 29
Boot light 138
Brake fluid level 81
Braking system check 118
Breakdowns 47
Breakdowns, emergency kit 53
Broken windscreen, dealing with 46
Bulbs, renewal 131
   Boot light 138
   Courtesy light 138
   Direction indicator repeater light 135
   Direction indicator (front) 133
   Engine compartment light 138
   Foglight (front) 133, 134
   Foglight (rear) 137
   Headlight 132
   Luggage compartment light 138
   Number plate light (rear) 136
   Rear lights 135
   Reversing light 137
   Sidelight (front) 133
Buying and selling 9

C

Caravan, towing 58
Central door locking system 37
Changing a wheel 48
Check control warning system 21
Child safety 63
Choke linkage 108
Cigarette lighter 30
Clock 20
Clutch pedal adjustment 121
Contact breaker points 108
Controls and equipment 15
Conversion factors 164
Coolant, drain and renew 123
Coolant level 80
Coolant strength, checking 107
Cooling system, flushing 124
Courtesy light 138
Crime prevention 71

D

Dimensions and weights 12
Direction indicator (front) 133
Direction indicator repeater light 135
Distance tables 166
Distributor, clean 112
Door locks 36
Door mirror switch 29
Driveshaft check 118
Driving:
   Abroad 65
   In bad weather 56
   On motorways 57
   Safety 55

VAUXHALL ASTRA & BELMONT

INDEX

E

Economical driving 69
Electric door mirror switch 29
Electric windows..30
Electricity (mains) and electrical equipment,
precautions concerning 100
Emergencies...39
Engine compartment light 138
Engine oil and filter, renewal 106
Engine oil, disposal of................................100
Engine types, identifying............................101
Equipment..15
Exhaust system 120

F

Fanbelt 116
Fault finding 143
    Brakes..146
    Electrics...148
    Engine 143
    Gearbox, transmission and clutch............145
    Suspension and steering..........................147
Faults while driving 118
Fire hazards, precautions concerning 99
Fire in vehicle, dealing with..........................46
First aid 40
Foglight (front) 133, 134
Foglight switch..28
Folding roof (Convertible) 38
Fuel gauge 20
Fuel inlet filter...120
Fuel pump filter screen 108
Fumes, precautions 99
Fuses 139

G

Gearbox (manual)..23
Gearbox (manual) oil level 114
Glossary 149

H

Handbrake adjustment 122
Hazard flasher switch 28
Head restraints 34
Headlamp beam adjustment control 25
Headlight 132
Heated rear window switch 29
Heating 26
Hinges and locks 116
Horn..15, 16, 86

I

Identifying engines 101
Ignition HT voltage, precautions concerning....100
Ignition switch/steering lock.........................22
Instrument illumination control....................29
Instruments ....17 to 20
Interior, cleaning 129
Interior mirror...32

J

Jacking and vehicle support 100
Jacking and wheel changing.................48, 100
Jargon 149
Jump lead starting 53

L

Leaks (fluid), checking for 87
Level control system.....................................37
Lights, checking operation 86
Lights switch (exterior lights)........................24
Local radio frequencies 159
Locks...36
Lubricants and fluids.....................................73
Luggage compartment light 138

VAUXHALL ASTRA & BELMONT

INDEX

M

Mains electricity, precautions	100
Maintenance	89
Maintenance schedule	90
MOT, preparing for	141
Brakes	141
General	142
Lights	141
Seatbelts	142
Steering and suspension	141
Tyres and wheels	142
Motorway breakdowns	48
Motorway driving	57
Multi-function switch	24, 25

N

Number plate light (rear)	136

O

Oil and filter, renewal	106
Oil level	80
Oil pressure gauge	20

P

Plugs (spark)	75, 112
Points	108
Power steering fluid level	122
Power steering pump drivebelt	116
Punctures	48
Push starting	52

R

Radio frequencies	159
Rear foglight switch	28
Rear lights	135
Rear seats	35
Rear view mirror	32
Relays	140
Rev. counter	20
Reversing light	137
Road test	118
Routine maintenance	89

S

Safety (child)	63
Safety (driving)	55
Safety first!	98
Scratches	130
Seasonal servicing	125
Seat adjustment	33
Seat belts	32
Security	71
Selling	11
Service schedule	90
Service specifications	73
Service tasks	106 to 125
Servicing	89
Servicing notes	172
Sidelight (front)	133
Skid control	60
Soft top (Convertible)	38
Spare parts	101
Spark plug specifications	75
Spark plugs	75, 112
Specifications	73
Speedometer	20
Starting a car with a flat battery	52
Steering column adjustment	30
Steering fluid level	122
Steering lock	22
Steering pump drivebelt	116
Sunroof	31
Suspension and steering check	118
Suspension level control system	37
Switches:	
Electric door mirror	29
Front foglight	28
Hazard flasher (warning)	28
Heated rear window	29
Horn	15, 16
Ignition	22
Lights	24
Multi-function	24, 25
Rear foglight	28
Steering column	24, 25
Tailgate wash/wipe	25
Windscreen wash/wipe	25

VAUXHALL ASTRA & BELMONT

INDEX

T

Tachometer 20
Tailgate wash/wipe switch 25
Temperature gauge 20
The Astra & Belmont family 7
Throttle linkage 108
Tools 128
Tow starting 52
Towing 51
Towing 51, 58
Trailer, towing 58
Troubleshooting ....see Fault finding
Tyre pressures 76
Tyres, checking 82

U

Underbody checking and cleaning 117, 126
Underbonnet layout ....78, 79, 102 to 105

V

Vandalism, dealing with 46
Ventilation 26
Voltmeter 20

W

Warning lights 21
   Brake pad wear 22
   Brake fluid, low level 21
   Bulb failure 22
   Coolant level 22
   Direction indicator 21
   Engine control 21
   Handbrake 21
   Headlight main beam 21
   Ignition 21
   Manual choke 21
   Oil level 22
   Oil pressure 21
   Trailer direction indicator 21
   Washer fluid level 22
Wash/wipe switch 25
Washer fluid level 83
Weights 13
Wheel changing 48
Windscreen breakage, dealing with 46
Windscreen wash/wipe switch 25
Wipers and washers, checking 85
Women drivers, advice to 61

VAUXHALL ASTRA & BELMONT

SERVICING NOTES

Date	Action

VAUXHALL ASTRA & BELMONT

SERVICING NOTES

Date	Action

174 SERVICING NOTES

Date	Action

VAUXHALL ASTRA & BELMONT